Cell Culture and Upstream Processing

Cell Culture and Upstream Processing

Edited by

Michael Butler

Department of Microbiology, University of Manitoba, Manitoba, Canada

Taylor & Francis
Taylor & Francis Group

Published by:

Taylor & Francis Group

In US: 270 Madison Avenue
          New York, N Y 10016
In UK: 2 Park Square, Milton Park
          Abingdon, OX14 4RN

© 2007 by Taylor & Francis Group

ISBN: 978-0-415-39969-2
      0-415-39969-6

Cover images kindly supplied by Tharmala Tharmalingam and Carly Steinfield.

A catalog record for this book is available from the British Library.

Library of Congress Cataloging-in-Publication Data

Cell culture and upstream processing / edited by Michael Butler.
     p. ; cm.
     "From presentations at the ... IBC conferences as well as meetings of the European Society for Animal Cell Technology (ESACT) and Protein Expression in Animal Cells (PEACe)."
     Includes bibliographical references.
     ISBN 978-0-415-39969-2 (alk. paper)
1. Cell culture--Congresses. 2. Biochemical engineering--Congresses. 3. Biotechnology--Congresses.
I. Butler, M. (Michael), 1947-
     [DNLM: 1. Cell Culture Techniques--methods--Congresses. 2. Genetic Engineering--methods--Congresses.
QS 525 C394 2007]

TP248.25.C44C449 2007
660.6'3--dc22
     2007005195

Senior Editor: Elizabeth Owen
Editorial Assistant: Kirsty Lyons
Senior Production Editor: Simon Hill
Typeset by: Phoenix Photosetting, Chatham, Kent, UK
Printed by: The Cromwell Press

Printed on acid-free paper

10 9 8 7 6 5 4 3 2 1

Taylor & Francis Group, an informa business   Visit our web site at http://www.garlandscience.com

Contents

Media development

Glycosylated proteins

Contributors

Michael Butler, Department of Microbiology, University of Manitoba, Winnipeg, Manitoba, R3T 2N2, Canada

Marco Cacciuttolo, Medarex, Inc., Bloomsbury, New Jersey, USA

Trevor N. Collingwood, Sangamo BioSciences, Inc., Richmond, California, USA

Ray Field, Cambridge Antibody Technology, Cambridge, UK

Stephen F. Gorfien, Invitrogen Corporation, Grand Island, NY 14072, USA

Roy Jefferis, Immunity & Infection, University of Birmingham, B15 2TT, UK

John Joly, Genentech Inc., 1 DNA Way, South San Francisco, California, USA

Helen Y. Kim, Amgen Inc. Protein Science, Thousand Oaks, California, USA

Lynne Krummen, Genentech Inc., 1 DNA Way, South San Francisco, California, USA

Yanmei Lu, Genentech Inc., 1 DNA Way, South San Francisco, California, USA

Gloria Meng, Genentech Inc., 1 DNA Way, South San Francisco, California, USA

Gerald Nakamura, Genentech Inc., 1 DNA Way, South San Francisco, California, USA

Domingos Ng, Genentech Inc., 1 DNA Way, South San Francisco, California, USA

Thomas Potgieter, GlycoFi, Inc., 21 Lafayette Street, Lebanon, NH 03766, USA

Brad Snedecor, Genentech Inc., 1 DNA Way, South San Francisco, California, USA

Amy Shen, Genentech Inc., 1 DNA Way, South San Francisco, California, USA

Fyodor D. Urnov, Sangamo BioSciences, Inc., Richmond, California, USA

Stefan Wildt, GlycoFi, Inc., 21 Lafayette Street, Lebanon, NH 03766, USA

Abbreviations

ACF	Animal component-free
ADCC	Antibody-dependent cellular cytotoxicity
APC	Activated protein C
ASGPR	Asialoglycoprotein receptor
BHK	Baby hamster kidney
BSE	Bovine spongiform encephalopathy
BURs	Base-unpairing regions
CDC	Complement-dependent cytotoxicity
CDG	Congenital disorders of glycosylation
CDM	Chemically defined media
CDRs	Complementarity-determining regions
cGMP	Current good manufacturing practices
CHO	Chinese hamster ovary
CHO-K1SV	Chinese hamster ovary K1 suspension variant
CMOs	Contract manufacturing organizations
CMV	Cytomegalovirus
CTCF	CCCTC-binding factor; zinc finger protein
DHFR	Dihydrofolate reductase
DIG	Digoxigenin
DSBs	Double-strand DNA breaks
EDQM	European Directorate for the Quality of Medicines
EDTA	Ethylenediaminetetraacetic acid
ELISA	Enzyme-linked immunosorbent assay
EMEA	European Agency for the Evaluation of Medicinal Products
EP	European pharmacopoeia
ER	Endoplasmic reticulum
FBS	Fetal bovine serum
FDA	Food and Drug Administration
FRT	Flp recombination target
FTEs	Full-time equivalents
FUT	Fucose transferase
GalT	Galactosyl transferase
GBA	Glucocerebrosidase
G-CSF	Granulocyte-colony stimulating factor
GFP	Green fluorescent protein
GM-CSF	Granulocyte-macrophage colony stimulating factor
GMO	Genetically modified organism
GMP	Good manufacturing practice
GNT	N-Acetyl glutamine transferase
GS	Glutamine synthetase
HACA	Human anti-chimeric antibody
HAHA	Human anti-human antibody
HAMA	Human anti-mouse antibody
HCCF	Harvested cell culture fluid
HCIC	Hydrophobic charge interaction chromatographic
HDR	Homology-directed repair
HEK-293	Human embryonic kidney
HITs	Homologous intergenic tracts
HMGA	High mobility group A; family of non-histone chromosomal proteins
HNRNPA2/ CBX3	Heterologous nuclear protein A2/chromobox homologue 3
HP1	Heterochromatin protein-1
HPLC	High performance liquid chromatography
HR	Homologous recombination
HygTk	Hygromycin B phosphotransferase-thymidine kinase
ICRs	Imprinting control regions
IFNβ	Interferon beta
IgG	Immunoglobulin G
IRES	Internal ribosomal entry site
JP	Japanese pharmacopoeia
LTR	Long terminal repeat

MALDI-TOF	Matrix-assisted laser desorption/ ionization time-of-flight
MAR	Matrix attachment region
MBL	Mannan-binding lectin
MBLs	Mannose-binding lectins
MCB	Master cell bank
MR	Mannose receptor
MSX	Methionine sulfoximine
NANA	*N*-Acetylneuraminic acid
NK	Natural killer
NS0	Mouse myeloma
ORF	Open reading frame
ORI	Origin of replication
OST	Oligosaccharyltransferase
PCR	Polymerase chain reaction
PER-C6	Human retina-derived
PFM	Protein-free media
PK	Pharmacokinetic
PNGase F	Peptide-*N*-glycosidase F
PTMs	Post-translational modifications
QA	Quality assurance
rEPO	Recombinant erythropoietin
rMAbs	Recombinant monoclonal antibodies
RMCE	Recombination-mediated cassette exchange
RT-PCR	Reverse transcription polymerase chain reaction
SAR	Scaffold attachment region
SATB-1	Special AT-rich binding protein
SFM	Serum-free media
siRNA	Small inhibitory RNA
SOPs	Standard operating procedures
ST	Sialyl transferase
Su(Hw)	Suppressor of hairy wing
TK	Thymidine kinase
TOC	Total organic carbon
t-PA	Tissue-type plasminogen activator
TSEs	Transmissible spongiform encephalopathies
UCOE	Ubiquitous chromatin opening elements
USP	United States pharmacopeia
VEGF	Vascular endothelial growth factor
ZFNs	Zinc finger nucleases
Zw5	Zest-white-5

Preface

Cell culture-based bioprocesses have become a major industrial production platform for a novel range of biopharmaceutical products that are being used in the treatment of various human diseases. The purpose of this multi-authored book is to provide a review of advances in the area of mammalian cell technology that have been important for the recent development of this technology.

Although animal cell cultures have been important at a laboratory scale for most of the last 100 years, it was the initial need for human viral vaccines in the 1950s (particularly for poliomyelitis) that accelerated the design of large-scale bioprocesses for mammalian cells. These processes required the use of anchorage-dependent cells and the modern version of this viral vaccine technology currently employs microcarrier support systems that can be used in pseudo-suspension cultures designed in stirred tank bioreactors.

More recent interest in mammalian cell culture bioprocesses is associated with recombinant protein technology developed in the 1970s and 1980s. The first human therapeutic protein to be licensed from this technology in 1982 was recombinant insulin (Humulin from Genentech) but the relative structural simplicity of this molecule allowed its large scale production to be developed in *Escherichia coli,* which is fast-growing and robust compared to mammalian cells. It was soon realized that the subsequent targets for recombinant therapeutics were more complex and required the post-translational metabolic machinery only available in eukaryotic cells. At the present time there are over 40 licensed biopharmaceuticals produced from mammalian cell bioprocesses. These are defined as recombinant proteins, monoclonal antibodies and nucleic acid-based products. Since 1996 the chimeric and humanized monoclonal antibodies have dominated this group with several blockbuster products such as Rituxan, Remicade, Synagis, and Herceptin. These hybrid construct molecules are far less immunogenic than their murine counterparts and have extended serum half-lives.

There have been several identifiable advances in animal cell technology that have enabled efficient biomanufacture of these products. Advances in molecular biology have provided gene vector systems that allow high specific protein expression and minimization of the undesirable process of gene silencing that may occur in prolonged culture. Characterization of cellular metabolism and physiology has enabled the design of fed-batch and perfusion bioreactor processes with significant improvements in product yield, some of which are now approaching 5 g L^{-1}. However, there is also a vigorous debate on the relative merits of perfusion or fed-batch culture strategies for maximizing productivity in these systems. Many of these processes are now being designed in serum-free and animal component-free media to ensure that products are not contaminated with the adventitious agents found in bovine serum. The therapeutic efficacy of many of these biopharmaceuticals is dependent upon molecular structure, which is often defined by the post-translational modification of the protein, such as glycosylation. Specific glycosylation profiles of a protein have been found to give enhanced therapeutic activity. These quality issues can be controlled by using selectively modified host cells and careful control of the culture process.

In this book, these key developments for the recent advances in the production of biopharmaceuticals from mammalian cell culture processes are discussed by prominent industrial and academic leaders in the field. The topics and authors have been selected from the best elements of oral presentations recently delivered by the authors at major international conferences that include a series of IBC conferences under the title 'Cell Culture and Upstream Processing', meetings of the European Society for Animal Cell Technology (ESACT) and Protein Expression in Animal Cells (PEACe). The authors of each chapter have expanded their oral presentations to provide a comprehensive review under the remit of their title, emphasizing generic elements of the area that would appeal to a wide scientific audience. It is hoped that this will provide useful information for both scientific practitioners of animal cell technology as well as students of biochemical engineering or biotechnology in graduate or high level undergraduate courses at university.

M. Butler
University of Manitoba,
Winnipeg, Canada
December, 2006

Overview on mammalian cell culture

Cell line development and culture strategies: future prospects to improve yields

1

Michael Butler

1.1 Introduction

Over the last few years there has been a rapid increase in the number and demand for approved biopharmaceuticals produced from animal cell culture processes. In part, this has been due to the development of several humanized monoclonal antibodies that have proved efficacious for the treatment of various human diseases and particularly different forms of cancer. *Figure 1.1* shows the increase in the number of biopharmaceuticals approved for production by mammalian cell bioprocesses over 20 years

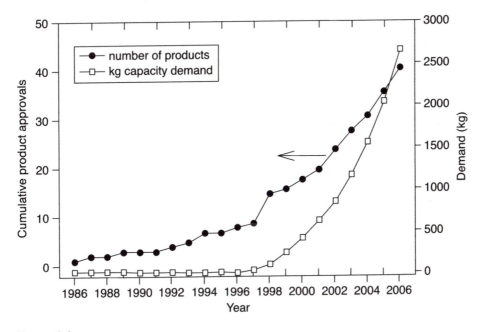

Figure 1.1

The demand for mammalian cell culture products. (Data taken from Molowa and Mazanet, 2003.)

with the actual and expected demand in kilograms. The sharp increase in the year 2000 is largely due to the number of approved therapeutic antibodies that are required at relatively high clinical doses.

At present there are up to 40 licensed biopharmaceuticals produced from mammalian cell bioprocesses (Molowa and Mazanet, 2003; Pavlou, 2003; Walsh, 2003). These are recombinant proteins, monoclonal antibodies and nucleic acid-based products.

Since 1996 monoclonal antibodies have dominated this group with up to 20 now licensed as therapeutics. These include such blockbuster products as Rituxan®, Remicade®, Synagis® and Herceptin® (Brekke and Sandie, 2003; Pavlou, 2004). The reason for the increased production of monoclonal antibodies is that the problem of the immunogenicity of the original murine-based antibodies has been solved allowing possibilities for their use as therapeutics. Chimeric antibodies (e.g. Rituxan®) are molecular constructs with reduced immunogenicity. These consist of a mouse variable region linked to the human constant region. A further step to humanizing an antibody can be made by replacement of the murine framework region, leaving only the complementarity determining regions (CDRs) that are of murine origin. These genetically constructed hybrid molecules are far less immunogenic than their murine counterparts and have serum half-lives of up to 20 days.

The commercial significance of these products as therapeutics is driving a vigorous interest for progress in process development using mammalian cell cultures. The high clinical dosage requirements for antibodies leads to large volume demands in biomanufacture (estimated to be >2000 kg for all biotherapeutic proteins). Because of the high demand for individual antibodies, the cost efficiency in manufacture becomes extremely important.

Mammalian cell culture bioprocesses gained prominence as a tool for the production of recombinant protein in the 1970s and 1980s. The first human therapeutic protein to be licensed from this technology in 1982 was recombinant insulin (Humulin® from Genentech) but the relative structural simplicity of this molecule allowed its large scale production to be developed in *Escherichia coli*, which is fast-growing and robust compared to mammalian cells. However, it was soon realized that the subsequent targets for recombinant therapeutics were more complex and required the post-translational metabolic machinery only available in eukaryotic cells.

There have also been several identifiable advances in animal cell technology that have enabled efficient biomanufacture of these products. Gene vector systems allow high specific protein expression and some minimize the undesirable process of gene silencing that may occur in prolonged culture. Characterization of cellular metabolism and physiology has enabled the design of fed-batch and perfusion bioreactor processes that has allowed a significant improvement in product yield, some of which are now approaching 5 g L^{-1}. Many of these processes are now being designed in serum-free and animal component-free media to ensure that products are not contaminated with the adventitious agents found in bovine serum.

In this chapter, some of the recent achievements and future prospects for the production of biopharmaceuticals from animal cell culture processes are discussed.

1.2 Cell line transfection and selection

The ability to produce and select a high producing animal cell line is key to the initial stages of the development of a cell culture bioprocess (Andersen and Krummen, 2002; Wurm, 2004). Chinese hamster ovary (CHO) cells have become the standard mammalian host cells used in the production of recombinant proteins, although the mouse myeloma (NS0), baby hamster kidney (BHK), murine C127 cells, human embryonic kidney (HEK-293) or human retina-derived (PER-C6) cells are alternatives. All these cell lines have been adapted to grow in suspension culture and are well-suited for scale-up in stirred tank bioreactors. The advantage of CHO and NSO cells is that there are well-characterized platform technologies that allow for transfection, amplification and selection of high producer clones. Transfection of cells with the target gene along with an amplifiable gene such as dihydrofolate reductase (DHFR) or glutamine synthetase (GS) have offered effective platforms for expression of the required proteins. *Figure 1.2* shows the reactions catalyzed by these two enzymes. In these systems, selective pressure can be applied to the cell culture with an inhibitor of the DHFR or GS enzymes that causes an increase in the number of copies of the transfected genes including the target gene.

The DHFR system is routinely used with CHO cells deficient in the dihydrofolate reductase activity (DHFR⁻). The target gene is delivered to the cells along with the DHFR marker gene, usually on the same plasmid vector (Gasser *et al.*, 1982; Lucas *et al.*, 1996). The expression vector normally contains a strong viral promoter to drive transcription of the recombinant gene and this is delivered into the cells by one of a number of possible non-viral transfer techniques. These include calcium phosphate, electroporation, lipofection or polymer-mediated gene transfer. The transfected cells are selected in media requiring the activity of DHFR for nucleotide synthesis and cell growth. Exposure of the cells to several rounds of gradually increased concentrations of the DHFR enzyme inhibitor, methotrexate promotes amplification of the DHFR and the co-transfected target gene. Methotrexate treatment enhances specific protein production following an increased gene copy number that can be up to several hundred in selected cells.

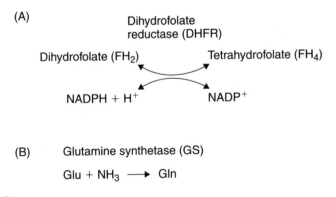

(A)

Dihydrofolate
reductase (DHFR)

Dihydrofolate (FH₂)   Tetrahydrofolate (FH₄)

NADPH + H⁺   NADP⁺

(B)   Glutamine synthetase (GS)

Glu + NH₃ ⟶ Gln

Figure 1.2

Enzymes associated with the most commonly used expression systems.

The glutamine synthetase (GS) expression system is an alternative that works as a dominant selectable marker, which is an advantage because this does not require the use of specific mutant cells (Bebbington *et al.*, 1992). NS0 cells are the preferred target cells because of the absence of an endogenous GS enzyme. In GS-transfected cells the enzyme allows the synthesis of glutamine intracellularly and so the transfected cells are selected in a glutamine-free culture media. The added advantage of this is that the cell cultures produce less ammonia, which is a potentially toxic metabolic by-product of mammalian cells that affects protein glycosylation and may inhibit cell growth. CHO cells are also used with this expression system but in this case methionine sulfoximine (MSX) is generally used as an inhibitor of the endogenous enzyme activity. In either case only cells containing the transfected GS and product genes are able to survive.

Gene amplification in this system can be mediated by MSX, which is required at concentrations of 10–100 μM to provide clones with amplified genes and sufficiently high specific productivities. For NS0 cells without endogenous GS activity, a lower activity of an enzyme inhibitor can be used for selection and amplification in NS0. Typically copy numbers of only four to ten genes per cell are found in these cells but they give as high expression levels as the cells from the DHFR system. The advantage of this is that the GS high producer clones can be produced in around 3 months, which is half the time it takes for the selection of DHFR clones.

High yields of recombinant proteins can also be produced from a human cell line – notably PER.C6, which was created by immortalizing healthy human embryonic retina cells with the E1 gene of adenovirus (Jones *et al.*, 2003). This cell line has been well characterized and has been shown to be able to produce high levels of recombinant protein with relatively low gene copy numbers and without the need for amplification protocols. The added value of these cells is that they ensure the recombinant proteins produced receive a human profile of glycosylation.

1.3 Increase in efficiency in selecting a producer cell line

The speed of production of early phase clinical material is critical in drug discovery and rapid screening. Cell line creation is generally the slowest step, taking up to 18 months in many cases. There are clear advantages for decreasing these timelines. Birch (2005) describes how the traditional timeline of a nominal 100 weeks for creation of a cell line can be reduced to 40 weeks by strategic modifications in protocols. *Figure 1.3* indicates how this can be done by the elimination of gene amplification (stage 2), a decrease in the time of cell cloning (stage 4) and the use of cells pre-adapted for growth in suspension (stage 5).

Gene amplification can be avoided through a number of strategies. Cells can be selected that have target genes integrated into highly efficient sites and high throughput screening may ensure the selection of highly productive cells. The integration of the expression vector in the host cell genome is critical for production of a high producer clone. The site of integration has a major effect on the transcription rate of the recombinant gene. Chapters 2 and 3 describe methods for targeting and protecting integrated genes.

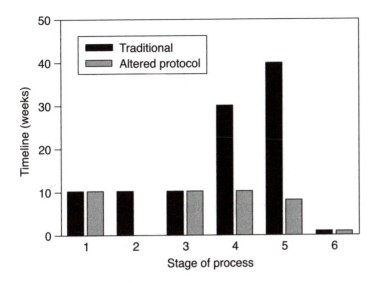

Figure 1.3

Timelines for constructing and selecting high quality clonal cells (from Birch, 2005). Stages of the process: (1) transfection and selection; (2) amplification and selection; (3) suspension evaluation; (4) cell cloning; (5) suspension evaluation; and (6) selection of lead cell lines.

One method of selecting high cell producers is to construct an expression vector with a weak promoter for the marker but a strong promoter for the product gene. This strategy has been used successfully with the GS expression system (Birch, 2005). The weak promoter for the GS gene increases the selection stringency for variants with a rare integration into transcriptionally efficient sites of the cellular genome. By using a strong promoter for the product gene in the same vector, expression of the product protein is enhanced at the favorable integration site.

A cell clone with a specific productivity of up to 10 pg/cell/day can be produced fairly routinely for recombinant protein production. However, higher specific productivities (up to 90 pg cell^{-1} day^{-1}) may be possible with improvements in vector technology and further understanding of the parameters that control protein expression in the cell (Wurm, 2004).

Screening for a high producer clone can be a lengthy process that depends upon assaying the secreted proteins to determine productivities of all candidate clones. The traditional way of doing this is by multiple rounds of limiting dilution of a cell population to ensure a high probability of clonality. This is time consuming (stage 4, *Figure 1.3*). Alternative methods based on high throughput selection may be devised to go more rapidly from transfection to a cloning stage. This may involve rapid enrichment of cell pools prior to cloning. These may be based upon rapid product assays or the use of flow cytometry to identify clones that have an appropriate product marker on the cell surface (Borth *et al.*, 2001; Carroll and Al-Rubeai, 2004). An alternative may be the method of capillary-aided cloning that offers a single-step protocol (Clarke and Spier, 1980). With this method small droplets (1 μL) are drawn into a capillary tube and deposited on a microwell plate. Visual confirmation ensures that the drops contain single cells.

Pre-selecting cells for high growth rates under bioreactor conditions may also decrease the development time of biomanufacture (stage 5 in *Figure 1.3*). This will avoid an extra stage of adaptation of the transfected cells for maximum growth. One example of this is the use of the CHO-K1 variant (CHO-K1SV) that has been adapted for growth in suspension in chemically defined medium (Birch, 2005). These cells are used as a host for the GS expression system that ensures high specific productivity, whilst the properties of the host cells ensure high yields with sustained viability at stationary phase in culture.

1.4 Stability of gene expression

The stability of selected clones over long term culture is a critical parameter for commercial production (Kim *et al.*, 1998). The application of selective pressure such as methotrexate in the case of the DHFR selection system causes gene amplification but a proportion of these genes are unstable and removal of the selective agent, as is necessary in production cultures, results in a gradual loss of the gene copy number. Fann *et al.* (2000) reported the step-wise adaptation of t-PA producing CHO cell lines to 5 μM MTX, which resulted in a maximum specific recombinant protein production of 43 pg cell^{-1} day^{-1} but on removal of the MTX the maximum productivity decreased to 12 pg cell^{-1} day^{-1} within 40 days. This decrease in productivity could be correlated with a reduction of gene copy number for individual clones.

Barnes *et al.* (2004) studied the stability of antibody expression from NS0 cells amplified with the GS system. They reported that there was a loss of mRNA for the recombinant protein over long-term culture but this was only reflected in a decrease in protein expression if the mRNA was below a threshold level. This indicates that selection of clones for high levels of recombinant mRNA may be useful as a predictor of stable protein production. Above a saturation level of mRNA it is argued that the limitation to protein expression resides in the translational/secretory machinery of the cell.

Progressive gene silencing can occur over successive culture passages following clonal isolation. This is thought to be associated with the spread of heterochromatin structure which is condensed and transcriptionally silent. Most expression systems cause random transgene integration into the host cell and this leads to positional effects that cause variable expression and stability. However, the genetic control elements that are responsible for establishing a transcriptionally active transgene are not fully understood. The negative positional effects of transgene insertion can be overcome by protecting the gene from silencing (as described in Chapter 2) or by targeted insertion into specific 'hot spots' of the cellular genome (as described in Chapter 3). Vectors in which the genes are flanked with insulators, boundary elements or ubiquitous chromatin opening elements may promote stable expression by insulating the transgene from positional effects of the chromatin. These elements that can be incorporated into expression vectors include matrix (or scaffold) attachment regions (MAR or

SAR) that allow an open chromatin structure to be maintained (Kim *et al*, 2004). This can allow higher efficiency of expression of the integrated genes (see Chapter 2). The ubiquitous chromatin opening elements (UCOEs) have also been incorporated into transgene vectors to prevent gene silencing and give consistent, stable and high level gene expression irrespective of the chromosomal integration site (Haines *et al.*, 2004).

1.5 Optimization of the fermentation process

The demand for higher quantities of biotherapeutics and particularly monoclonal antibodies are met by increased volumetric sizes of bioreactors. However, the product yield can also be improved significantly by the design of media and feeding strategies that can maintain high concentrations of viable cells for prolonged periods of time.

A producer cell clone may be grown in a batch culture to above 10^6 cells mL^{-1} over 3–4 days to allow synthesis and product secretion. The limits for growth and production are related to the accumulation of metabolic by-products (such as ammonia and lactate) or the depletion of nutrients such as glucose or glutamine (Butler and Jenkins, 1989). It has been shown for some time that the growth of cells can be extended through perfusion culture where the constant supply of nutrients and the removal of media can lead to cell densities of at least 10^7 cells mL^{-1} (Butler *et al.*, 1983). These principles have been used to develop fed-batch cultures, which have been shown to be operationally simple, reliable and flexible for multi-purpose facilities (Bibila and Robinson, 1995; Cruz *et al.*, 2000; Xie and Wang, 1997). The most successful strategies involve feeding concentrates of nutrients based upon the predicted requirements of the cells for growth and production. This can involve slow feeding of low concentrations of key nutrients. The maintenance of low concentration set-points of the major carbon substrates enables a more efficient primary metabolism which leads to lower rates of production of metabolic by-products, such as ammonia and lactate. As a result the cells remain in a productive state over extended time-frames. The strategic use of fed-batch cultures has enabled considerable enhancement of yields from these processes. This is often combined with a biphasic strategy of production in which cell proliferation is allowed in the first phase so that high cell densities accumulate, followed by a phase in which cell division is arrested to allow cells to attain a high specific productivity. In this type of strategy growth can be arrested by a decrease in culture temperature (Fox *et al.*, 2004; Yoon *et al.*, 2003). By directly supplying cells with a balanced nutrient feed, a fed-batch culture can now be expected to yield upwards of 2 g L^{-1} of recombinant protein, which is probably at least 10-fold higher than the maximum that could be expected by a simple batch culture in standard culture medium.

The increased efficiency of mammalian cell bioprocesses over the last two decades is illustrated in *Figure 1.4*, which shows typical profiles of culture processes operated in 1986 versus 2004. A traditional batch process reaches a maximum cell density of 3×10^6 cells mL^{-1} after 100 h followed by a gradual decline in viability caused by deteriorating conditions in the culture (*Figure 1.4A*). However, the strategy of using a well-designed fed-batch process allows higher maximum densities (up to 10^7 cells mL^{-1}) to be

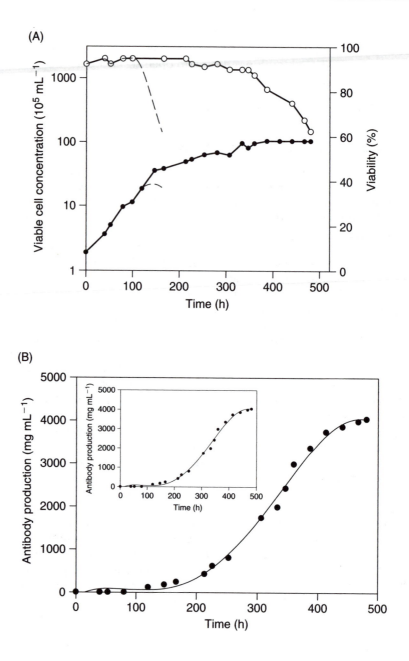

Figure 1.4

Comparison of typical profiles from cell culture processes from 1986 and 2004 are shown. (A) The viable cell concentration and viability of cells in culture is shown for two processes taken from data in 1986 (—) and 2004 (● ○). (B) The accumulated monoclonal antibody production for 2004 is shown in the main graph with the values from the 1986 process shown in the insert. (The graphs have been redrawn from data from Lonza Biologics as published in Wurm, 2004, with permission from the Nature Publishing Group.)

achieved with a sustained high viability up to 400 h. *Figure 1.4B* shows the effect of the improved cell concentration profile on accumulated antibody production. In this case the final yield of the 2004 process leads to antibody concentrations around 4 g L^{-1}, which is approximately two orders of magnitude greater than the 1986 process.

1.6 Apoptosis

The prolonged viability of cells in the fed-batch process is brought about by preventing apoptosis, which is a form of programmed cell death regulated through a cellular cascade of activities in response to one of a number of factors such as nutrient depletion, metabolic by-product accumulation, excessive shear forces or hypoxia (Arden and Betenbaugh, 2004). Characteristic cellular changes include DNA fragmentation, chromatin shrinkage followed by membrane blebbing and the formation of apoptotic bodies. Nutrient feeding can provide protection and this is normally used as the first preventive measure to control the cellular environment to delay apoptosis. Serum is known to contain unidentified anti-apoptotic factors that can offer protection (Zanghi *et al.*, 1999). However, the serum-free formulations that are required for production processes make the cells more vulnerable to apoptosis. Some supplements such as suramin or insulin growth factor may provide independent anti-apoptotic protection in serum-free cultures (Sunstrom *et al.*, 2000; Zanghi *et al.*, 2000). There are also other specific caspase inhibitors available to suppress apoptosis (Tinto *et al.*, 2002) but their expense in large-scale cultures is likely to be prohibitive. A number of anti-apoptotic genes (e.g. *bcl-2* or *bcl-x$_2$*) have been identified and may be transfected into a host cell. These inhibit the release of pro-apoptotic molecules from the mitochondria and may prolong the viability of the cell. This strategy has been shown to work for several cell lines, which show higher viabilities and improved robustness under conditions that would normally cause apoptosis (Kim and Lee, 2002; Mastrangelo *et al.*, 2000; Tey *et al.*, 2000).

1.7 Bioreactors

Animal cell cultures are normally grown in stainless-steel stirred-tank bioreactors that are designed with impellers that minimize shear forces (Kretzmer, 2002). Producer cells can be made to be sufficiently robust in this environment if they are provided with suitable growth media and gas sparging is carefully controlled. The capacity of commercial bioreactors for animal cells has gradually increased over the past two decades – with capacities now reported up to 20 000 L from some of the larger biopharmaceutical companies. Air-lift bioreactors have also been applied to large-scale animal cells and these have been shown to be efficient for protein production.

Perfusion cultures are more demanding to set up on a large scale but they have the potential advantage of allowing a continuous stream of product over several weeks or even months (Mercille *et al.*, 2000). A further advantage is the rapid removal of any potentially labile products from the culture environment. An effective cell separator will allow the protein-containing media to be fed directly into a chromatography column suitable for extraction and downstream processing (Castilho and Medronho, 2002; Shirgaonkar *et al.*, 2004; Wen *et al.*, 2000). A further advantage of this mode of culture is that the bioreactor may be up to ×10 smaller for the production of the same quantity of product (Ryll *et al.*, 2000).

This area of bioprocess design will become of even greater importance as some of the first generation blockbuster drugs (e.g. erythropoietin, human growth hormone and α-interferon) start being produced as generics (Walsh, 2003). Eleven biopharmaceuticals with combined annual sales of $13.5 billion lose patent protection in 2006 (Walsh, 2003). The challenge then will be to produce bioequivalents in efficient low cost bioprocesses.

1.8 The capacity crunch

With an increase in the number and demand for recombinant biopharmaceuticals, there is a requirement for greater biomanufacturing capacity. This created a major problem in 2001 when the demand for Enbrel, a recombinant fusion protein commercialized by Immunex for the treatment of rheumatoid arthritis exceeded expectations. However, there was insufficient large-scale culture manufacturing capacity to meet this clinical demand – even by contract manufacturers available at that time. Although the problem of Enbrel was eventually solved, the episode highlighted a general problem that the pipeline of biotherapeutic products is expanding more rapidly than the world capacity for cell culture production (Mallik *et al.*, 2002). It is estimated that the present world capacity for cell culture production stands at 475 000 L of which 75% is controlled by biopharmaceutical companies and the remainder by contract manufacturers. Since 2001 there has been a substantial increase in biomanufacturing capacity to meet this demand, but some estimates suggest that there is still a potential shortfall with manufacturing demand continuing to exceed the production capacity (Molowa and Mazanet, 2003). The estimates of the shortfall vary and depend upon a number of difficult predictions (Thiel, 2004).

One of the reasons for this extra demand for biomanufacturing capacity is the dose requirement for the novel therapeutic humanized monoclonal antibodies that are now being commercialized. The requirement for hundreds of kilograms per annum far exceeds other recombinant therapeutics such as erythropoietin which is more potent at smaller doses. The extra demand for production is being met by the construction of increased bioreactor capacity by some biopharmaceutical companies such as Biogen, Lonza Biologics and Genentech. However, the requirement for large capacity bioreactors may be off-set by an increased productivity of cell culture systems, as illustrated in *Figure 1.4* with present systems capable of producing concentrations up to 5 g L^{-1}, which is up to 100× the productivity that would have been expected a few years ago. Clearly, an enhancement of cell line productivity reduces the volumetric capacity required of the bioreactor for manufacture by an equivalent factor.

A rider to this problem of the capacity crunch is the personnel crunch, with a shortage of highly qualified personnel available to manage the impending demand for production of the new series of cell culture products. It is clear that the number of graduates or PhDs with expertise in the key areas of this technology is declining at a time when the demand for such expertise is likely to be expanding (Mallik *et al.*, 2002). It is hoped that this problem is recognized so that the full potential of animal cell technology for biopharmaceutical production can be achieved.

Acknowledgment

The author would like to thank the Natural Sciences and Engineering Research Council of Canada (NSERC) for financial support in animal cell technology.

References

Andersen DC and Krummen L (2002) Recombinant protein expression for therapeutic applications. *Curr Opin Biotechnol* **13**: 117–123.

Arden N and Betenbaugh MJ (2004) Life and death in mammalian cell culture: strategies for apoptosis inhibition. *Trends Biotechnol* **22**: 174–180.

Barnes LM, Bentley CM and Dickson AJ (2004) Molecular definition of predictive indicators of stable protein expression in recombinant NS0 myeloma cells. *Biotechnol Bioeng* **85**: 115–121.

Bebbington CR, Renner G, Thomson S, King D, Abrams D and Yarranton GT (1992) High-level expression of a recombinant antibody from myeloma cells using a glutamine synthetase gene as an amplifiable selectable marker. *Biotechnology (NY)* **10**: 169–175.

Bibila TA and Robinson DK (1995) In pursuit of the optimal fed-batch process for monoclonal antibody production. *Biotechnol Prog* **11**: 1–13.

Birch J (2005) *Upstream Mammalian Cell Processing – Challenges and Prospects.* Bioprocess International presentation, Berlin. Conference Handbook from www.ibclifesciences.com

Borth N, Zeyda M, Kunert R and Katinger H (2001) Efficient selection of high-producing subclones during gene amplification of recombinant Chinese hamster ovary cells by flow cytometry and cell sorting. *Biotechnol Bioeng* **71**: 266–273. Erratum in: *Biotechnol Bioeng* (2002) **77**: 118.

Brekke OH and Sandlie I (2003) Therapeutic antibodies for human diseases at the dawn of the twenty-first century. *Nat Rev Drug Discov* **2**: 52–62.

Butler M (2004) *Animal Cell Culture and Technology*, 2nd Edn. Bios Scientific, Oxford.

Butler M and Jenkins H (1989) Nutritional aspects of growth of animal cells in culture. *J Biotechnol* **12**: 97–110.

Butler M, Imamura T, Thomas J and Thilly WG (1983) High yields from microcarrier cultures by medium perfusion. *J Cell Sci* **61**: 351–363.

Carroll S and Al-Rubeai M (2004) The selection of high-producing cell lines using flow cytometry and cell sorting. *Expert Opin Biol Ther* **4**: 1821–1829.

Castilho LR and Medronho RA (2002) Cell retention devices for suspended-cell perfusion cultures. *Adv Biochem Eng Biotechnol* **74**: 129–169.

Clarke JB and Spier RE (1980) Variation in the susceptibility of BHK populations and cloned cell lines to three strains of foot-and-mouth disease virus. *Arch Virol* **63**: 1–9.

Cruz HJ, Moreira JL and Carrondo MJ (2000) Metabolically optimised BHK cell fed-batch cultures. *J Biotechnol* **80**: 109–118.

Fann CH, Guirgis F, Chen G, Lao MS and Piret JM (2000) Limitations to the amplification and stability of human tissue-type plasminogen activator expression by Chinese hamster ovary cells. *Biotechnol Bioeng* **69**: 204–212.

Fox SR, Patel UA, Yap MG and Wang DI (2004) Maximizing interferon-gamma production by Chinese hamster ovary cells through temperature shift optimization: experimental and modeling. *Biotechnol Bioeng* **85**: 177–184.

Gasser CS, Simonsen CC, Schilling JW and Schimke RT (1982) Expression of abbreviated mouse dihydrofolate reductase genes in cultured hamster cells. *Proc Natl Acad Sci USA* **79**: 6522–6526.

Haines AWS, Wedgwood J, Cliffe S, Simpson D, Mountain A and Irvine A (2004) Rapid antibody production from stable pools of mammalian cells and isolation of high expressing clones using UCOEs. Presentation at *Cell Culture and Upstream Processing*, IBC Life Sciences, Berlin. Conference Handbook from www.ibclifesciences.com

Jones D, Kroos N, Anema R *et al.* (2003) High-level expression of recombinant IgG in the human cell line per.c6. *Biotechnol Prog* **19**: 163–168.

Kim NS and Lee GM (2002) Response of recombinant Chinese hamster ovary cells to hyperosmotic pressure: effect of Bcl-2 overexpression. *J Biotechnol* **95**: 237–248.

Kim NS, Kim SJ and Lee GM (1998) Clonal variability within dihydrofolate reductase-mediated gene amplified Chinese hamster ovary cells: stability in the absence of selective pressure. *Biotechnol Bioeng* **60**: 679–688.

Kim JM, Kim JS, Park DH, Kang HS, Yoon J, Baek K and Yoon Y (2004) Improved recombinant gene expression in CHO cells using matrix attachment regions. *J Biotechnol* **107**: 95–105.

Kretzmer G (2002) Industrial processes with animal cells. *Appl Microbiol Biotechnol* **59**: 135–142.

Lucas BK, Giere LM, DeMarco RA, Shen A, Chisholm V and Crowley CW (1996) High-level production of recombinant proteins in CHO cells using a dicistronic DHFR intron expression vector. *Nucleic Acids Res* **24**: 1774–1779.

Mallik A, Pinkus GS and Sheffer S (2002) Biopharma's capacity crunch. *The McKinsey Quarterly* **9**.

Mastrangelo AJ, Hardwick JM, Zou S and Betenbaugh MJ (2000) Part II. Overexpression of bcl-2 family members enhances survival of mammalian cells in response to various culture insults. *Biotechnol Bioeng* **67**: 555–564.

Mercille S, Johnson M, Lanthier S, Kamen AA and Massie B (2000) Understanding factors that limit the productivity of suspension-based perfusion cultures operated at high medium renewal rates. *Biotechnol Bioeng* **67**: 435–450.

Molowa DT and Mazanet R (2003) The state of biopharmaceutical manufacturing. *Biotechnol Annu Rev* **9**: 285–302.

Pavlou AK (2003) Marketspace: Trends in biotherapeutics. *J Commercial Biotechnol* **9**: 358–363.

Pavlou AK (2004) The immunotherapies markets, 2003-2008. *J Commercial Biotechnol* **10**: 273–278.

Ryll T, Dutina G, Reyes A, Gunson J, Krummen L and Etcheverry T (2000) Performance of small-scale CHO perfusion cultures using an acoustic cell filtration device for cell retention: characterization of separation efficiency and impact of perfusion on product quality. *Biotechnol Bioeng* **69**: 440–449.

Shirgaonkar IZ, Lanthier S and Kamen A (2004) Acoustic cell filter: a proven cell retention technology for perfusion of animal cell cultures. *Biotechnol Adv* **22**: 433–444.

Sunstrom NA, Gay RD, Wong DC, Kitchen NA, DeBoer L and Gray PP (2000) Insulin-like growth factor-I and transferrin mediate growth and survival of Chinese hamster ovary cells. *Biotechnol Prog* **16**: 698–702.

Tey BT, Singh RP, Piredda L, Piacentini M and Al-Rubeai M (2000) Influence of bcl-2 on cell death during the cultivation of a Chinese hamster ovary cell line expressing a chimeric antibody. *Biotechnol Bioeng* **68**: 31–43.

Thiel KA (2004) Biomanufacturing, from bust to boom ... to bubble? *Nat Biotechnol* **22**: 1365–1372.

Cell line development and culture

Tinto A, Gabernet C, Vives J, Prats E, Cairo JJ, Cornudella L and Godia F (2002) The protection of hybridoma cells from apoptosis by caspase inhibition allows culture recovery when exposed to non-inducing conditions. *J Biotechnol* **95**: 205–214.

Walsh G (2003) Biopharmaceuticals benchmarks – 2003. *Nat Biotechnol* **21**: 865–887.

Wen ZY, Teng XW and Chen F. (2000) A novel perfusion system for animal cell cultures by two step sequential sedimentation. *J Biotechnol* **79**: 1–11.

Wurm FM (2004) Production of recombinant protein therapeutics in cultivated mammalian cells. *Nat Biotechnol* **22**: 1393–1398.

Xie L and Wang DI (1997) Integrated approaches to the design of media and feeding strategies for fed-batch cultures of animal cells. *Trends Biotechnol* **15**: 109–113.

Yoon SK, Kim SH and Lee GM. (2003) Effect of low culture temperature on specific productivity and transcription level of anti-4-1BB antibody in recombinant Chinese hamster ovary cells. *Biotechnol Prog* **19**: 1383–1386.

Zanghi JA, Fussenegger M and Bailey JE (1999) Serum protects protein-free competent Chinese hamster ovary cells against apoptosis induced by nutrient deprivation in batch culture. *Biotechnol Bioeng* **64**: 108–119.

Zanghi JA, Renner WA, Bailey JE and Fussenegger M (2000) The growth factor inhibitor suramin reduces apoptosis and cell aggregation in protein-free CHO cell batch cultures. *Biotechnol Prog* **16**: 319–325.

The producer cell line

Use of DNA insulator elements and scaffold/matrix-attached regions for enhanced recombinant protein expression

2

Helen Y. Kim

2.1 Introduction

As recombinant proteins continue to emerge as an important class of human therapeutics, improving methods to generate cell lines expressing high levels of desired proteins is becoming increasingly critical. The ability to generate high-expressing cell lines is important for the production of clinical candidate molecules as well as for rapid and reliable production of molecules for characterization and validation studies. In either case, the method should be rapid, cost-effective and scaleable. Specifically, a method that can generate a high expressing cell clone without the extensive selection and screening step will be useful for the production of a clinical candidate molecule, and a method that can rapidly and reliably generate proteins in the range of hundreds of milligrams would be equally valuable for early phase validation studies. For production of proteins in the range of hundreds of milligrams, using a pool or a collection of transfected cells, rather than a single cell clone, can reduce resource as well as the time it takes to obtain the protein, significantly. However, the widely variable expression levels of different cell clones within a transfection necessitates extensive single cell cloning and screening process, and prevents the use of transfected cell pools for production, even on a small scale. Furthermore, the expression levels of transfected pools typically decrease with culture time. This instability of pool expression is likely due to relatively low levels of recombinant protein expression by the majority of the cells within a given transfection, and subsequent selective survival of the low-expressing cells, since the low expressing cells often display growth advantage compared to the high expressing cells.

2.2 The position effect

Reasons underlying the large variability between different clonal expression levels include different plasmid copy numbers, and a phenomenon known as 'position effect', initially described in *Drosophila melanogaster* as position-effect variegation (Henikoff, 1992). Position effects are influences from the surrounding DNA which affect a gene's expression upon integration into the genome. These influences include activities of external enhancers and silencers, as well as heterochromatinization. In most cell line development processes, expression constructs are introduced into the genome of the host cell, using methods which result in random integration. Hence, the level of expression of the transgene will depend on the genomic environment where the DNA integrates. It is then, not surprising that a large majority of cell clones do not express high levels of the transgene, considering that most genomic sites are transcriptionally repressive (Festenstein *et al.*, 1996; Goetze *et al.*, 2005). These repressive effects can spread in cis, resulting in epigenetic silencing of adjacent genes. Transcriptional repression can occur due to histone deacetylation (Jeppesen *et al.*, 1992) and methylation, at lysine 9 of histone H3 (H3-K9) (Peters *et al.*, 2001), as well as methylation of the promoter sequence of the transfected DNA. These events can be regulated by the local availability of heterochromatin associated components such as heterochromatin-1 (HP1) (Eissenberg *et al.*, 1992; Festenstein *et al.*, 1999; reviewed in Maison, 2004) or histone-H3 methyltransferases, such as suppressor of variegation 3-9 in *D. melanogaster*, also known as Suv39h1 and Suv39h2 in the mouse (Rea *et al.*, 2000). HP1 proteins are multidomain proteins with several binding partners, which can function as structural adaptors for the assembly of macromolecular complexes in chromatin, leading to heterochromatin assembly and maintenance (Maison, 2004). HP1 interacting partners include the DNA methyltransferases Dnmt1 and Dnmt3a, which are involved in CpG methylation (Fuks *et al.*, 2003).

2.3 Use of insulators and S/MARs can reduce the effects of heterochromatin on transgene expression

There are two common approaches that can be used to protect DNAs from negative position effects, or the integration-dependent repressive effect on the transfected gene expression. One approach is to direct transgene integration into a predetermined site that is transcriptionally active, using site-specific recombination methods. The approach of gene insertion into pre-determined 'hot spots' is explained in Chapter 3. Another method is to simply incorporate DNA sequence elements found in chromatin border regions, into the expression vector such that regardless of the integration site, your gene will be protected from the surrounding chromatin influence. This chapter will focus on the latter method only, and summarize some of the recent developments in vector engineering methods, which helped to overcome the negative position effects of randomly integrated DNAs. This method takes advantage of nature's solution for protecting the transcriptionally active regions from epigenetic heterochromatinization. In order to do this, cells have evolved barrier mechanisms, using specialized DNA sequences that can establish chromatin borders, known as insulator or chromatin barrier elements (*Figure*

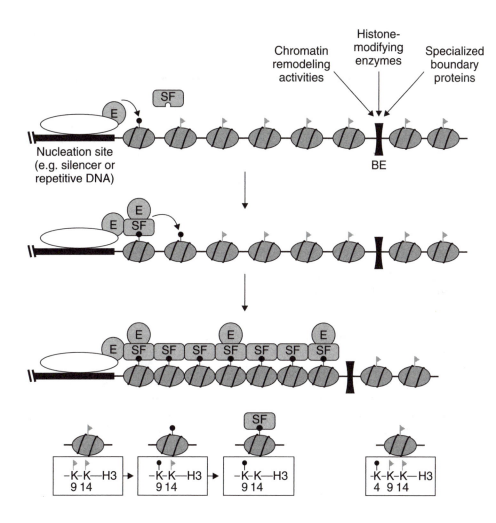

E, Histone-modifying enzyme

SF, Silencing factor

Figure 2.1

Model for formation of silenced chromatin domains. After the recruitment to a specific heterochromatin nucleation site by proteins that directly bind DNA or are targeted by way of RNAs, histone-modifying enzymes (E) such as deacetylases and methyltransferases modify histone tails to create a binding site for silencing factors (SF). After this nucleation step, self-association of silencing factors (such as Swi6/HP1 or Sir3) is hypothesized to provide an interface for their interaction with histone-modifying enzymes, which then modify adjacent histones, creating another binding site for silencing factors. Sequential rounds of modification and binding result in the stepwise spreading of silencing complexes along nucleosomal DNA for several kilobases (spreading). Spreading of silencing complexes is blocked by the presence of boundary elements (BE). The modifications associated with the amino terminus of histone H3 in fission yeast heterochromatin (bottom left) and euchromatin (bottom right) are illustrated as an example. Deacetylation and methylation of H3 Lys9 are followed by deacetylation of H3 Lys14 and create a binding site for the Swi6 silencing factor.

2.1, from Grewal and Moazed, 2003). For the purpose of recombinant protein expression, sequences that can behave as chromatin borders and protect the transfected gene from the surrounding chromatin influences include insulator sequences as well as scaffold/matrix –attachment regions (S/MARs). Expression studies from my laboratory as well as several other laboratories, using such elements, have shown that flanking a transgene with insulators or S/MARs can suppress the clonal expression variability (Girod *et al.*, 2005, Goetze *et al.*, 2005). Generation of higher proportion of transfected cells with improved expression levels can lead to reduced number of clones that need to be screened to identify an acceptable production cell line. This further enables the generation of high-expressing transfected pools with improved stability, potentially eliminating the need to isolate clones for production of research material for early characterization and validation studies.

2.4 DNA insulator elements

DNA insulator elements were initially identified in *Drosophila*, as sequences that prevent external enhancers from inappropriately activating the promoter of a reporter gene (Dorsett, 1993; Geyer and Corces, 1992; Kellum and Schedl, 1991). These are DNA sequences which, in their natural state, are thought to define distinct, chromatin domains of gene expression. Such elements prevent cross-regulation of adjacent genes or gene clusters by restricting the activity of DNA elements such as enhancers and silencers to the domain in which they reside. There are several such DNA elements identified to date from various species, including *Saccharomyces cerevisiae*, *D. melanogaster*, *Xenopus lavis*, sea urchin, chicken, and human, and all are found between independently regulated gene loci (*Figure 2.2*; Bell *et al.*, 2001). Many of the above insulator elements have been shown to protect recombinant genes from position effects *in vivo*, across species, as well as *in vitro*, in mammalian cell lines, suggesting that these elements have a conserved role in defining domains of gene expression (Bell *et al.*, 2001). Consistent with a function in defining the domains of gene expression, insulator elements are found throughout the eukaryotic genomes (Kuhn and Geyer, 2003). Subsequent analyses of these insulator elements have revealed a wide diversity of insulator sequences, and suggested that many insulators are compound elements, containing several distinct protein binding sites, and separable properties (Recillas-Targa *et al.*, 2002). The 1.2-kb DNA sequence element (5'HS4) at the 5' end of the chicken β-globin locus, for example, can be separated to enhancer blocking and barrier activities. Results from such experiments have helped to define two main functions for insulators: the ability to block enhancer–promoter activity (enhancer blocking activity), and the ability to prevent the spread of heterochromatin (barrier activity). For the purpose of expression vector engineering, experiments which delineate distinct activities of insulator elements may help to reduce the size of effective elements to be used in the expression vector, thus making the final size of the expression plasmid easier to manage. Furthermore, the potential benefit on expression would be obvious, if expression vectors can be constructed with sequences which possess the barrier activity only, without the enhancer blocking activity.

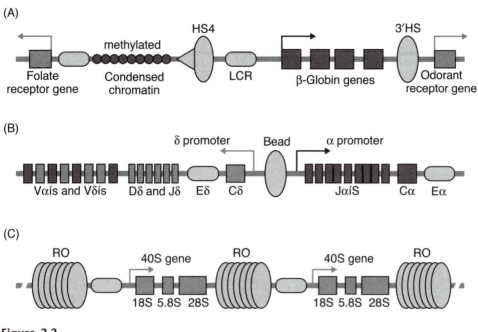

Figure 2.2

Vertebrate insulators (ovals) are found between genes and enhancers with distinct profiles of expression. (A) Hypersensitive sites (HS) found in the chicken β-globin locus. (B) The blocking element alpha/delta (BEAD) localized in the human TCR α/δ locus. (C) Repeat organizer (RO) localized within the tandem arrays of *Xenopus* ribosomal RNA genes.

2.5 The scaffold/matrix-attachment regions

The scaffold/matrix-attachment regions (S/MARs) are experimentally identified sequences that are associated with the nuclear scaffold or matrix, and thought to be responsible for the attachment of chromatin loops to the nuclear scaffold or matrix (Laemmli *et al.*, 1992; Michalowski *et al.*, 1999). These sequence elements are further thought to be involved in chromatin remodeling and subsequent transcriptional activation, as well as protection of transgenes from position effects (Allen *et al.*, 2000; Bode *et al.*, 2000). Similar to the sequence diversity found in insulators, S/MAR sequences tend to be heterogeneous. However, S/MAR sequences have been analyzed more extensively, and subsequent identification of frequently occurring sequence motifs found in large numbers of experimentally characterized S/MARs, led to the development of several computational tools to predict yet unidentified S/MAR elements (Benham *et al.*, 1997; Frisch *et al.*, 2002; Glazko *et al.*, 2001; Singh *et al.*, 1997). S/MARs, as well as insulators, can be identified on the basis of their position in the genome in different species, suggesting evolutionary conservation (Avramova *et al.*, 1998; Greally *et al.*, 1999; Whitehurst *et al.*, 1992). *Figure 2.3*, taken from a recent publication by Glazko *et al.* (2003), shows two examples of sequence conservation between orthologous human and mouse intragenic sequences, in which the human sequences are derived from experimentally characterized S/MARs.

(A)

```
HIT1
Human TTTCTTATTGTAACCATCCTGGAGTGTGTGAGGTGGTACCTCATTGTGGTTTTGATTTGCATTTCCCTAATGACTAATGATGTTGAGCATCTTTT
mouse TGTCTTAGTGTGACCATCCTCCTGTTTAT-------------AAACTGGCTAGGACTTACATTGTCCTAATGATGACTGGTGGTGAA--ATCTACA
      * ***** *** ******** **   ** **        *   *** *  ** ** ****  ******** * * **** *    ****

Human CATGTATTTGTTGGCCATTTTTGTATCTTCTTTG-28-ATTCTTAAAATTGGATTGTTTGTCTTTTTATTATTGAATTGCAAGATTCTT--TAT
Mouse CA--CTCTGCTTGGTCATTTCGGTTTCCTCCTGG    ACTCTTTAAAATTAGTTGTT--TCTTCTTATTATTGCATTGTAAAAATTCAAAATCT
      ** *  **** ***** ** ** ** * *        * **** *** *  ***** **** ********* **** ** * ***      *

Human GTATTCTACATGC-------AAGACTCATTAGAT
Mouse GAATTCTTGGGGCTGGAGAGAAGGCTCAGTGGTT
      * *****     **      *** **** * * *

HIT2
Human TAAGTTATATGAAAGTGCTT---TGAAAATATCACATTGAAGGAATTTATTTTAAATGGGATTCA---CATGTTGAGGAACGATTGTTCAATTTT
Mouse TGAGTAAAAGGGAAGTGCCTGCCTTCAACTTCCAAATTATGGGGTGTTATTTTCAACAAGATTTATTTCCTGTGAAGGAATGATCATTAAATTTT
      * *** * * ******* *     * ** ** *** **   ** ******* **   **** *  *  * **  ***** *** ** ** ******

Human TACA -92-CCTGGGAAAAATAGCGAATTCCAGGTCTAGAGCAGGAAATG-TGCCATCTTGTCACGCCAGACAGCAAAGGAACCATCAAGGCC-A
Mouse AAGA-245-CCAAGGAAAGATGGCTACTTCCAGATCCATAAAAAGAAATGACACCATCACATCA-----CACAGTCAGAAAGCCATCAAGAGCAA
      * *      ** ***** **  *  ** ****** **   * ** ****  ***** *   ***      ***** *  * ********* *  *

Human ACCGGGTCTGTGACAAAAGGACTCAGAAGACA  -9-GACTCCTACT-GACCAAAAATAGGACAATGTGAGCCTTAATAAAGATAATAATTGCAAA
Mouse ATTAGGGCTTGAACAAAATGGATCTGAAGACA-22-GGCTCCCACTAGCTAAAAAATGGGACAATTTGAGCATCAATCTAAATAATAA-TACAAA
      *  ** **  ****** ** ** ********  * **** *** *   ****** ******** ***** * ***  * ******* * ****

Human GAATTGAAAAATAACAACTATGCTCAAATCCATGAGT--ATAAATGATAGTAAAATAAA
Mouse GCATTAGAAAGTATTAATGGGGCTCCAATCCATGTGTTGATAGAAGAAACTAAAAAGGA
      * *** *** ** **     **** ******** ** *** * ** * *****    *
```

TRENDS in Genetics

(B)

```
HIT1
Human TATATTTATGTGTTCATAAACA-ATATGTAATTTTACAATATGTAACACACTAGTAACATACTAA----------TTTAAAACTTGTTTTTAGTT
Mouse TATATACATGTGTGTATATATATATATGTATATATA-AATTTATAAAAAATTA-TAAAATTATAAGAAGAGATACTTCATAATTTTTTATTTTATT
      ****  ******  *** * * *******  *** *  * ** *  * **  ** ***   ***          ** * ** ** * *** **

Human TACAAATATGTT -21-ACATATTGTACAACCTATTGTAAAATAAAAACAG -64-TTGATTTTGTGGTTTTA---TTATTATCTTAACCTACACT
Mouse TATTTTTTGTT-540-ATAGATATTTCAAACTATTTCAAAATAGAATTAG-338-TTGATATTGTTATTTAAGGCTTATTACATTTACCTCCAAT
      **  * ****     ***  **** ** *** ***** **** **  * **       ***** **** *     ****** * **** ** *

Human -TTTAATTAATCCATTTTATTTGTTACATGGTATTCTATTATATCATAAAA
Mouse GATAATGTATTCCATTTTACTAGTT-TTGAGTATTATATTATATAATTATA
       * *  ** ********* * *** *     ***** ******** ** * *

HIT2
Human TATTCTGTTGGTTTTTGTTGTTGGTCATTTGAGACCAT
Mouse TATTCTGTTGGTTTTTGTTTGT-TTGATTTGAAAATAT
      ******************* *  *  * ****** *  **

HIT3
Human ATTCAAGTGGTTTTGGTGCCTCAGCCTCCTGAGTAGCTGGGATTATAGGCGTGTGCCACCA---TGCCCAGCTAATTTTTGTATTTTT
Mouse ACTCAAGAGGTCCACCTGCTTCTGTCTCCCTAATA-CTGGAATTAAAGGTGTACGCTACCACTATGCCTGACTTTTTTTTTTTTTTTTT
      * ***** ***    *** ** * **** *  ** * **** ****  *** *** ***  *     ***   ** ***** * *****

HIT4
Human CGGGATTTCACCATGTTGGCCAGGGTGGTCTTGAACT -7-AGTGATCTGCTCACCTCAGCTGCACAAAGTGCTGGGATTACAGGTGTTAGCCAC
Mouse CTGGAACTCACTTTGTAGACCAGGCTGGCCTCGAACT-37-AGAAATCCGCCTGCCTCTGCCTC-CTGAGTGCT-GGATTAAGGGTGT--GCCAC
      * ***  **** ****** * ***** *** ** ****    **  *** ** ** *** ***  * ***** ** **** * * ***   *****

Human CATACCCTGCCTCTATTCTCTTGTTAAGA-44-CTGGAAGAAACACA-TGTTTCTTGTGTATATGAATGAAAATTTGTTTTATACATTAGATATT
Mouse CACGCCCGGCTTATTCCCTCTTTATAAAA -1-CTTTAAGAAAGAAATAATACTTCTTTCTTACAATGTAATTTAGATGTATAT-TTACATTTT
      ** *** ** *  ** * ***** ** *    ** * *****  * * **  ** ****   * * ***  ** ** * ** *** ***

Human TCCAAATTGTTCT-149-TGCATTGGGATGAAGTTGAAACCTAATATATTTTCCAAATGAGTAAA
Mouse GTCATCTTATTGT-458-TTCAGTGTGATGAAGCCATTCCCTCTTATTTTCTCCTAATGTCCAAA
      **  ** ** *        * ** ** *******   ***  *** ** ** *** ****   ***
```

TRENDS in Genetics

Figure 2.3

Local similarities (HITs) in experimentally identified human matrix-scaffold attachment regions (MARs) and orthologous mouse regions for the (A) CSP-B gene (M62717) flanking sequence and (B) human topoisomerase I gene (L23999).

Analyses of experimentally identified S/MARs revealed a typical element to be as short as 300 base pairs to several kilobases in length. These S/MARs may contain several sequence motifs, including the base-unpairing regions (BURs) which are AT-rich nucleotide motifs, thought to function as DNA-unwinding elements (Kohwi-Shigematsu and Kohwi, 1990). Other sequences are: kinked DNA generated by the presence of TG, CA or TA dinucleotides separated from each other by 2–4 or 9–12 nucleotides; potential replication origins (ORIs) and homeotic protein-binding sites; intrinsically curved DNA produced by the $(A)_n$, T_mA_n; transcription factor binding sites; triple-helical or H-DNA structure sequences; and retroelement insertion hotspots (Avramova et al., 1998; Bode et al., 1996; Paul and Ferl, 1998; Tikhonov et al., 2001). In general, S/MARs contain approximately 70% AT. S/MARs are enriched with all the motifs listed above, but each individual S/MAR does not necessarily contain each motif. When incorporated into expression vectors, some S/MARs can function as insulators, by protecting against repressive effects of the neighboring chromatin environment, albeit to different extents (Goetze et al., 2005; Mlynarova et al., 2003). This is not surprising, since insulators and S/MARs can share similar sequence motifs (data not shown).

2.6 Binding proteins for DNA insulators and S/MARs

Proteins that bind to insulators and S/MAR elements have been identified, and subsequent analyses revealed these are DNA binding proteins that can modulate chromatin structure, alter histone acetylation or DNA methylation. These include the classical insulator protein, CCCTC binding factor (CTCF) and the Gli-type zinc-finger protein, YY1, and the S/MAR binding proteins, special AT-rich binding protein (SATB-1), and the high mobility group (HMGA) family of proteins. The 11 zinc-finger protein CTCF was originally identified as a protein which binds to the chicken β-globin 5′ insulator sequence, and subsequent studies have shown that a single CTCF-binding site is necessary and sufficient for enhancer blocking properties of the chicken β-globin 5′ insulator (Bell et al., 1999). CTCF is thought to regulate gene expression by affecting local genomic methylation characteristics by binding to its site and excluding DNA methyltransferases from the region. CTCF binding itself appears to be sensitive to methylation; therefore, the lack of CTCF binding can result in default methylation. CTCF is also thought to be involved in establishing higher order chromatin structures which affects gene expression (for a review, see Lewis and Murrell, 2004). Sequence analysis in my group indicates the presence of CTCF binding sites in other vertebrate insulators, as well as in some S/MARs (data not shown). Interestingly, several studies have shown that, depending on the context of the binding site, CTCF can also act as a transcriptional activator (Ohlsson et al., 2001). Another insulator-binding protein, YY1, was identified in the mouse Peg3 gene insulator (Kim et al., 2003). Similar to CTCF, YY1 is also a zinc-finger protein, displays methylation-sensitive binding, and involved in genomic imprinting, by regulating methylation patterns of imprinting control regions (ICRs) (Bell and Felsenfeld, 2000; Fedoriw et al., 2004; Kim et al., 2003). Imprinting refers to parent-origin-specific gene expression, where the parent-specific allelic repression is regulated by ICRs. Several studies have demonstrated that some ICRs can act as insulators and prevent promoter–enhancer interactions. In addition to CTCF and YY1, several insulator-binding proteins have been identified in *D. melanogaster*. These proteins include the protein suppressor of hairy wing [Su(Hw)], zest-white-5 (Zw5), and BEAF-32 (for review, see Bell, 2001).

SATB-1 was originally cloned based on its binding to a consensus core unwinding element derived from the BUR motif (Dickinson *et al.*, 1992). SATB-1 recognizes and binds a special AT-rich sequence context where one strand consists of mixed A's, T's, and C's, excluding G's (ATC sequences). Interestingly, SATB1 appears to bind along the minor groove with very little contact with the bases, suggesting that SATB1 recognizes a structure motif, rather than a specific sequence. Similarly, several other S/MAR-binding proteins also display a 'relaxed' sequence specificity (Bode *et al.*, 2000). SATB-1 regulates gene expression by regulating the location of genomic regions, with respect to the bases of chromatin loop domains, as well as affecting local histone modification states (Cai *et al.*, 2003; Dickinson *et al.*, 1992). HMGA (fka HMG-I/Y) proteins are nonhistone chromatin proteins involved in diverse cellular processes, including gene expression and DNA replication, recombination, and repair (Bustin, 1999). This is a class of proteins containing a DNA-binding domain, known as the AT-hook (Bustin, 2001). The HMGA proteins, using the AT-hook, recognize the local structure of AT-rich regions of the minor groove, resembling the mechanism of SATB-1 binding of chromatin (Reeves, 2001). In this way, one of the functions of the HMGA1 proteins is to behave as architectural transcription factors, thus regulating gene expression by modulating chromatin structure. Other MAR-binding proteins include the ubiquitously expressed SAF-A, Cux/CDP, and MeCP2 (Romig *et al.*, 1992; Scheuermann and Chen, 1989; Weitzel *et al.* 1997), as well as the tissue-specific protein, Bright, expressed specifically in activated B cells (Herrscher *et al.*, 1995).

2.7 DNA insulators or S/MARs can be incorporated into expression vectors

A simple method to increase recombinant protein expression is to take advantage of the insulating activities of S/MAR or insulator sequences, by incorporating the sequence element directly into the expression vector, such that the transgene can integrate into the host cell genome along with the protective sequences. The association of a S/MAR or an insulator with a transgene can lead to the expression cassette forming its own active chromatin domain, and may allow gene expression based on the strength of the promoter and enhancer included in the vector, rather than based on the transcriptional activity of the integration environment. This strategy has been used successfully both in mammalian as well as in plant systems (Allen *et al.*, 1996; Girod *et al.*, 2005; Kim *et al.*, 2004; McKnight *et al.*, 1992; Mlynarova *et al.*, 1995; Phi-Van *et al.*, 1990; Zahn-Zabal *et al.*, 2001). Some studies also show copy-number dependent expression when insulators or S/MARs have been used (McKnight *et al.*, 1992; Phi-Van *et al.*, 1990). Results from our studies and published studies show that the protective activities of either insulators or S/MARs can be orientation-dependent (*Figure 2.4*; Kanduri *et al.*, 2002; Kim *et al.*, 2003).

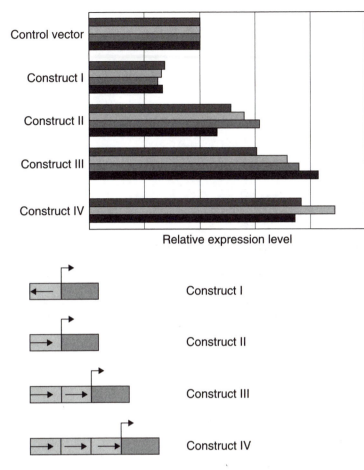

Figure 2.4

Analysis of stable CHO transfection pools demonstrate different copies of S/MAR and the orientation of S/MAR can affect final expression levels of a secreted protein. The constructs contain one (constructs I and II), two (construct III), or three copies (construct IV) of S/MAR element at the 5′ border of the promoter driving a secreted recombinant protein. Constructs I and II have the S/MAR element in reverse orientation. Results are from three or four separate transfections, performed in triplicates using LF2000 (Invitrogen), and the expression levels were determined by ELISA. Control vector contains a distinct insulator element at the same position in the expression vector.

Studies published by Zahn-Zabal *et al.* (2001) and Girod *et al.* (2005) demonstrated that adding the MAR element on a distinct plasmid by *trans*, in addition to the expression vector (*cis*) can increase recombinant gene expression up to 10-fold above control. Observations in my laboratory also demonstrated that addition of either insulators or S/MARs either in *cis* only, or in *cis* and in *trans*, can increase recombinant gene expression. We observed increased gene expression of the total transfected cell population, and this increase appeared to correlate with increased proportion of cell clones that express medium or high levels of the test gene (*Figure 2.5*). In

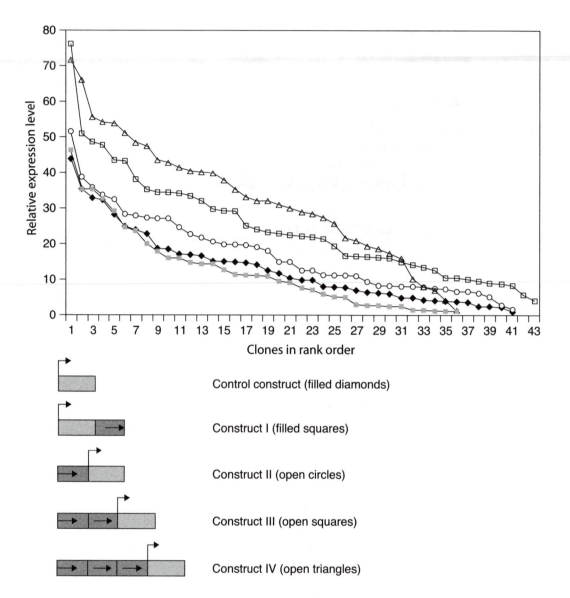

Figure 2.5

Expression analysis of stably transfected CHO cell clones show different copies of S/MAR, and the position of S/MAR respect to the transcript start site, can affect final expression levels of a secreted protein. The constructs contain one (constructs I and II), two (construct III), or three copies (construct IV) of S/MAR element at the 5' border (constructs II, III, and IV) or at the 3' border (construct I) of the promoter driving a secreted recombinant protein. Results are from clones combined from two separate transfections, using LF2000 (Invitrogen). Expression levels were determined by ELISA. Control vector contains a distinct insulator element at the same position, at the 5' border, in the expression vector.

addition to the general shift in the expression profiles of individual clones, using constructs with two or three copies of S/MAR in *cis* generated exceptionally high expressing cell clones (*Figure 2.5*). Additional benefit of using two copies of S/MAR versus one copy of S/MAR was more evident when cells from the transfected pools were maintained in culture over time. CHO cells, stably transfected with GFP, exhibited increased stability in expression, when kept in culture for 10 weeks post-transfection (data not shown). Using three copies of S/MAR in *cis* did not produce the exceptionally high expressing clones, nor a significant shift of expression profiles of individual clones, compared to the clones derived from transfections using two copies of S/MAR in *cis*. We think this 'plateau' of benefit in expression with three copies of S/MAR may attribute to the significant increase in plasmid size and a potential reduction in transfection efficiency. Consistent with this interpretation, we have seen increased pool expression from *cis* and *trans* transfections, which presumably introduces multiple copies of S/MAR. Additionally, although we have not carried out extensive evaluation of cell clones derived from *cis* and *trans* transfections, it seems reasonable to expect proportionally more high expressing cell clones in such cases. Similarly, studies by Girod *et al.* (2005) reported a general increase in the proportion of medium/high producer cells and a decrease of low producer cells within a transfected pool of cells. These data are consistent with the insulating role of insulators and S/MARs, especially if we assume that most genomic sites are transcriptionally repressive. Girod *et al.* (2005) also reported the appearance of a new 'very high' GFP- expressing population of cells, using the S/MARs. We have not seen similar subpopulation in our GFP transfections, although this may be due to properties of different GFPs as well as different gating parameters, since we readily observe appearance of significantly higher expressing cell clones when secreted test genes are used.

The insulator or S/MAR element can be added to a vector, in *cis*, either at the 5' of the promoter/enhancer or at the 3' of the polyA sequence, or both. Recent work published by Goetze *et al.* (2005), using singly integrated reporter plasmids, demonstrated an additive benefit on expression when S/MARs are present at both the 5' and the 3' borders of the reporter gene, possibly from providing protection of genomic influences from both sides of the expression cassette. It is conceivable that having the insulator or the S/MAR element at one end of the expression cassette may suffice when transfection methods allowing multiple-copy integration events are used, since each plasmid within the integration locus should contain the insulator or the S/MAR element, creating multiple mini domains of active gene expression. Using experimental methods allowing multiple-copy integration, and tandem arrays of the transgene within the integration locus, we have, in some cases, but not always, observed additional protection when these genomic elements are added at both ends (data not shown). One possible interpretation for our results may be that multiple copies of insulators and S/MARs provide additional protection, therefore, this may be a copy-number dependent additive phenomenon. This explanation is consistent with results from my group which showed additive benefit when we incorporated multimers of an insulator or an S/MAR element in the expression vector (*Figures 2.4 and 2.5*). Moreover, additional results from my group, as well as results by others, demonstrated increased expression level of recombinant proteins, when an additional S/MAR is introduced using the *trans* method (Girod, 2005). Therefore, additional copies of insulators or S/MARs can be introduced either by *trans*, or simply by incorporating multiple copies into the expression vector itself. In our experience, having the insulator or the S/MARs built into the expression vector produces more consistent results than introducing additional copies by *trans* alone. However, incorporating multiple copies into the expression vector can pose new challenges since this can quickly increase the size of the expression vector. Published studies suggest that the large DNA size can greatly reduce the gene cloning efficiency, and more importantly, it can also greatly reduce the transfection efficiency, potentially leading to reduced expression. This reduced transfection efficiency with large plasmids was observed for both supercoiled and linearized DNAs (Kreiss *et al.*, 1999). Some reports also suggested decreased efficiency of DNA entry into the nucleus when large plasmids are used. This, however, would be more relevant in the cases of nondividing cells, where the nuclear entry may depend on nuclear pores, but less relevant when the host cell is rapidly dividing, such as the CHO or the HEK 293 cells. In rapidly dividing cells, the entry into the nucleus is thought to occur during cell division when the nuclear envelope is already disrupted. In the interest of keeping the vector size relatively small, the use of both *cis* and *trans* method in combination can be beneficial (Girod *et al.*, 2005).

2.8 DNA insulators and S/MARs act in a context-dependent manner

Incorporating the insulator or the S/MAR element into the expression cassette can help to improve gene expression significantly but there are several parameters that need to be optimized in order to fully benefit from this technology. In addition to insulators or S/MARs, factors that can influence the expression vector strength may depend on several components, such as the promoter, enhancer, polyA, as well as the selection cassette. Subsequently, the final vector effectiveness depends not on any individual component, but rather on cross-talk and interplay among different components in the vector. In some cases, even the distances between different components appear to be important. Accordingly, the activities of insulator or S/MAR elements can be highly influenced depending on the context of the entire vector. Experiments performed in my group have demonstrated that certain combinations of promoters and insulator or S/MARs work more effectively than others in improving gene expression. We have seen examples where the different bacterial backbone of the plasmid DNA also seems to contribute to the final vector effectiveness. *Figure 2.5* shows an example of a vector context-dependent activity of a S/MAR element: transfections using a vector with S/MAR at the 5′ border of the expression cassette (construct II) produced significantly higher expressing CHO cell clones, compared to transfections using a vector with S/MAR at the 3′ border of the expression cassette (construct I). Examples from published studies include experiments using retroviral vectors and S/MARs derived from the human IFNβ gene locus. In this study, the S/MARs strongly supported transcription when placed at a distance of approximately 4 kb from the transcription initiation site, whereas the S/MARs almost completely shut off transcription when placed at a distance of approximately 2.5 kb (Shübeler *et al.*, 1996). These results clearly demonstrate the importance of distance and the resulting three-dimensional relationship between the S/MARs and other regulatory elements in the vector. In addition to the vector context, the different genomic context also influences the effectiveness of different insulators and the S/MARs, as shown in the studies published by Goetze *et al.* (2005). These context dependent phenomena are not surprising, considering the complexity of gene regulation *in vivo*. Transcription complexes regulating the eukaryotic polymerase type II promoters are highly ordered structures, and several lines of studies have demonstrated that changing the order, orientation or distances between the transcription factor binding sites can have a profound effect on gene expression.

Several studies have suggested a possible role of S/MARs in increased recombinogenic potential of a transgene, based on the observation that higher transgene copy numbers are obtained when S/MARs are included in the transfection experiment (Bode *et al.*, 1996; Girod *et al.*, 2005; Kim *et al.*, 2004). Increased recombinogenic potential may facilitate processes such as methotrexate-induced gene amplification, however, in theory, may also lead to decreased long-term stability. To date, there are no published reports suggesting either increased methotrexate-induced gene amplification, or decreased long-term stability due to specific S/MARs. We have not tested increased amplifibility of insulators or S/MARs to any great extent, and we have not detected instability in gene expression, using such elements.

2.9 Conclusion

In addition to insulators and S/MARs described in this chapter, other classes of genomic elements have been identified and successfully used in generating improved stable cell lines. The ubiquitous chromatin opening elements (UCOEs), identified from the heterologous nuclear protein A2/chromobox homologue 3 (HNRNPA2/CBX3) locus, have been used in combination with the human CMV promoter to generate high expressing cell lines (Antoniou *et al.*, 2003; Benton *et al.*, 2002). The antirepressor elements were identified from a human genomic DNA library, based on their ability to block gene silencing mediated by polycomb group (PcG) proteins, HP1 and HPC2 (Kwaks *et al.*, 2003). Similar to insulators and S/MARs, the anti-repressor elements are also highly conserved between human and mouse sequences (Kwaks *et al.*, 2003).

Another class of noncoding DNA, conserved between human and mouse intergenic regions, are the homologous intergenic tracts (HITs) (Glazko *et al.*, 2003). Although the functions of these HITs are less clear than other genomic elements described in this review, sequence analysis has revealed approximately 11% of HITs overlapped with predicted S/MARs, and conversely, more than 50% of predicted S/MAR sequences overlapped with HITs, suggesting potential function of HITs in chromatin and gene regulation. At this time, there are no published reports on using specific HITs in expression vectors, to improve gene expression.

Finally, it is worth noting that it may be possible to identify similarly high expressing cell clones, using an expression vector without S/MARs, or insulators, if one screens enough clones. Given the labor intensive nature of clone screening and selection procedures, it would be a significant benefit regardless, if we can reduce timeline and resource requirements by identifying the optimal cell clone by screening a few hundred clones rather than several thousand clones. In addition, increased transfected pool expression can completely eliminate the need to isolate clones, for producing recombinant proteins for early discovery research purposes. Many early discovery research experiments require proteins in the range of a few milligrams, for *in vitro* studies, to hundreds of millligrams for *in vivo* validation studies. In some cases, transient expression platforms may be used effectively, especially if the particular recombinant protein expresses well. In our experience, stable cell line approach proved to be more beneficial, either if the particular recombinant protein is difficult to express, or if there is a likely need of additional material in the future. In such cases, being able to use a transfected pool of cells, rather than isolating single cell clones, can save resources, and shorten the timelines significantly. Furthermore, information from experiments that delineate the enhancer-blocking activities from the barrier activities of insulators, can provide valuable information for future vector engineering efforts. It may be possible to generate expression vectors that can prevent the spread of heterochromatin in the integration site, but allow enhancer activities to further increase the expression of your recombinant gene. Lastly, one should keep in mind that gene expression is a multi-step process where transcription is only the beginning. Post- transcriptional modification and translation, followed by folding, post-translational modification and targeting are additional processes that can profoundly influence the final yield of a recombinant protein, and each process presents its unique challenge that is beyond the scope of this review.

Acknowledgments

I would like to thank Jennitte Stevens and Tom Boone for helpful discussions and critical reading of this review. I am grateful to Amgen Protein Science Department, especially Frank Martin and Tom Boone, for supporting my work for the past several years. Finally, I would like to give special thanks to the people in my group, both past and present, for their dedication and contribution.

An abbreviated version of this chapter first appeared in a May 2006 supplement to *BioProcess International* (www.bioprocessintl.com) titled 'Cell line engineering: improving bioproduction through science.'

References

Allen GC, Hall Jr G, Michalowski S, Newman W, Spiker S, Weissinger AK and Thompson WF (1996) High-level transgene expression in plant cells: effects of a strong scaffold attachment region from tobacco. *Plant Cell* **8**: 899–913.
Allen GC, Spiker S and Thompson WF (2000) Use of matrix attachment regions (MARs) to minimize transgene silencing. *Plant Mol Biol* **43**: 361–376.

Antoniou M, Harland L, Mustoe T *et al.* (2003) Transgenes encompassing dual-promoter CpG islands from the human TBP and HNRPA2B1 loci are resistant to heterochromatin-mediated silencing. *Genomics* **82**: 269–279.

Avramova Z, Tikhonov A, Chen M and Bennetzen JL (1998) Matrix attachment regions and structural colinearity in the genomes of two grass species. *Nucleic Acids Res* **26**: 761–767.

Bell AC and Felsenfeld, G (2000) Methylation of a CTCF-dependent boundary controls imprinted expression of the *Igf2* gene. *Nature* **405**: 482–485.

Bell AC, West AG and Felsenfeld G (1999) The protein CTCF is required for the enhancer blocking activity of vertebrate insulators. *Cell* **98**: 387–396.

Bell AC, West AG and Felsenfeld G (2001) Insulators and boundaries: versatile regulatory elements in the eukaryotic genome. *Science* **291**: 447–450.

Benham C, Kohwi-Shigematsu T and Bode J (1997) Stress-induced duplex DNA destabilization in scaffold/matrix attachment regions. *J Mol Biol* **274**: 181–196.

Benton T, Chen T, McEntee M, Fox B, King D, Crombie R, Thomas TC and Bebbington C (2002) The use of UCOE vectors in combination with a preadapted serum free, suspension cell line allows for rapid production of large quantities of protein. *Cytotechnology* **38**: 43–46.

Bode J, Stengert-Iber M, Kay V, Schlake T and Dietz-Pfeilstetter A (1996) Scaffold/matrix-attached regions: topological switches with multiple regulatory functions. *Crit Rev Eukaryot Gene Expr* **6**: 115–138.

Bode J, Benham C, Knopp A and Mielke C (2000) Transcriptional augmentation: modulation of gene expression by scaffold/matrix-attached regions (S/MAR elements). *Crit Rev Eukaryot Gene Expr* **10**: 73–90.

Bustin M (1999) Regulation of DNA-dependent activities by the functional motifs of the high-mobility-group chromosomal proteins. *Mol Cell Biol* **19**: 5237–5246.

Bustin M (2001) Revised nomenclature for high mobility group (HMG) chromosomal proteins. *Trends Biochem Sci* **26**: 152–153.

Cai S, Han HJ and Kohwi-Shigematsu T (2003) Tissue-specific nuclear architecture and gene expression regulated by SATB1. *Nature Genet* **34**: 42–51.

Dickinson LA, Joh T, Kohwi Y and Kohwi-Shigematsu T (1992) A tissue-specific MAR/SAR DNA-binding protein with unusual binding site recognition. *Cell* **70**: 631–645.

Dorsett D (1993) Distance-independent inactivation of an enhancer by the suppressor of hairy-wing DNA-binding protein of *Drosophila*. *Genetics* **134**: 1135–1144.

Eissenberg JC, Morris GD, Reuter G and Hartnett T (1992) The heterochromatin-associated protein HP-1 is an essential protein in *Drosophila* with dosage-dependent effects on position-effect variegation. *Genetics* **131**: 345–352.

Fedoriw AM, Stein P, Svoboda P, Schultz R and Bartolomei MS (2004) Transgenic RNAi reveals essential function for CTCF in *H19* imprinting. *Science* **303**: 238–240.

Festenstein R, Tolaini M, Corbella P, Mamalaki C, Parrington J, Fox M, Miliou A, Jones M and Kioussis D (1996) Locus control region function and heterochromatin-induced position effect variegation. *Science* **271**: 1123–1125.

Festenstein R, Sharghi-Namini S, Fox M, Roderick K, Tolaini M, Norton T, Saveliev A, Kioussis D and Singh P (1999) Heterochromatin protein 1 modifies mammalian PEV in a dose- and chromosomal-context-dependent manner. *Nature Genet* **23**: 457–461.

Frisch M, Frech K, Klingenhoff A, Cartharius K, Liebich I and Werner T (2002) *In silico* prediction of scaffold/matrix attachment regions in large genomic sequences. *Genome Res* **12**: 349–354.

Fuks F, Hurd PJ, Deplus R and Kouzarides T (2003) The DNA methyltransferases associate with HP1 and the SUV39H1 histone methyltransferase. *Nucleic Acids Res* **31**: 2305–2312.

Geyer PK and Corces VG (1992) DNA position-specific repression of transcription by a *Drosophila* zinc finger protein. *Genes Dev* 6: 1865–1873.

Girod PA, Zahn-Zabal M and Mermod N (2005) Use of the chicken lysozyme 5 matrix attachment region to generate high producer CHO cell lines. *Biotechnol Bioeng* 91: 1–11.

Glazko GV, Rogozin IB and Glazkov MV (2001) Comparative study and prediction of DNA fragments associated with various elements of the nuclear matrix. *Biochim Biophys Acta* 1517:351–356.

Glazko GV, Koonin EV, Rogozin IB and Shabalina SA (2003) A significant fraction of conserved noncoding DNA in human and mouse consists of predicted matrix attachment regions. *Trends Genet* 19: 119–124.

Goetze S, Baer A, Winkelmann S, Nehlsen K, Seibler J, Maass K and Bode J (2005) Performance of genomic bordering elements at predefined genomic loci. *Mol Cell Biol* 25: 2260–2272.

Greally JM, Gray TA, Gabriel JM, Song L, Zemel S and Nicholls RD (1999) Conserved characteristics of heterochromatin-forming DNA at the 15q11–q13 imprinting center. *Proc Natl Acad Sci USA* 96: 14430–14435.

Grewal SIS and Moazed D (2003) Heterochromatin and epigenetic control of gene expression. *Science* 301: 798–802.

Henikoff S (1992) Position effect and related phenomena. *Curr Opin Genet Dev* 2: 907–912.

Herrscher RF, Kaplan MH, Lelsz DL, Das C, Scheuermann R and Tucker PW (1995) The immunoglobulin heavy-chain matrix-associating regions are bound by Bright: AB cell-specific trans-activator that describes a new DNA-binding protein family. *Genes Dev* 9: 3067–3082.

Jeppesen P, Mitchell A, Turner B and Perry P (1992) Antibodies to defined histone epitopes reveal variations in chromatin conformation and underacetylation of centric heterochromatin in human metaphase chromosomes. *Chromosoma* 101: 322–332.

Kanduri C, Fitzpatrick G, Mukhopadhyay R, Kanduri M, Lobanenkov V, Higgins M and Ohlsson R (2002) A differentially methylated imprinting control region within the Kcnq1 locus harbors a methylation-sensitive chromatin insulator. *J Biol Chem* 277: 18106–18110.

Kellum R and Schedl P (1991) A position-effect assay for boundaries of higher order chromosomal domains. *Cell* 64: 941–950.

Kim J, Kollhoff A, Bergmann A and Stubbs L (2003) Methylation-sensitive binding of transcription factor YY1 to an insulator sequence within the paternally expressed imprinted gene, Peg3. *Hum Mol Genet* 12: 233–245.

Kim JM, Kim JS, Park DH, Kang HS, Yon J, Baek K and Yoon Y (2004) Improved recombinant gene expression in CHO cells using matrix attachment regions. *J Biotechnol* 108: 95–105.

Kohwi-Shigematsu T and Kohwi Y (1990) Torsional stress stabilizes extended base unpairing in suppressor sites flanking immunoglobulin heavy chain enhancer. *Biochemistry* 29: 9551–9560.

Kreiss P, Cameron B, Rangara R, Mailhe P, Aguerre-Charriol O, Airiau M, Scherman D, Crouzet J and Pitard B (1999) Plasmid DNA size does not affect the physico-chemical properties of lipoplexes but modulates gene transfer efficiency. *Nucleic Acids Res* 27: 3792–3798.

Kuhn EJ and Geyer PK (2003) Genomic insulators: connecting properties to mechanism. [Review]. *Curr Opin Cell Biol* 15: 259–265.

Kwaks TH, Barnett P, Hemrika W *et al.* (2003) Identification of anti-repressor elements that confer high and stable protein production in mammalian cells. *Nat Biotechnol* 21: 553–558.

Laemmli UK, Käs E, Poljak L and Adachi Y (1992) Scaffold-associated regions: cis-acting determinants of chromatin structural loops and functional domains. *Curr Opin Genet Dev* **2**: 275–285.

Lewis A and Murrell A (2004) Genomic imprinting: CTCF protects the boundaries. *Curr Biol* **14**: 284–286.

Maison C and Almouzni G (2004) HP1 and the dynamics of heterochromatin maintenance. [Review]. *Nat Rev Mol Cell Biol* **5**: 296–304.

McKnight RA, Shamay A, Sankaran L, Wall RJ and Hennighausen L (1992) Matrix-attachment regions can impart position-independent regulation of a tissue-specific gene in transgenic mice. *Proc Natl Acad Sci USA* **89**: 6943–6947.

Michalowski SM, Allen GC, Hall Jr GE, Thompson WF and Spiker S (1999) Characterization of randomly-obtained matrix attachment regions (MARs) from higher plants. *Biochemistry* **38**: 12795–12804.

Mlynarova L, Jansen RC, Conner AJ, Stiekema WJ and Nap JP (1995) The MAR-mediated reduction in position effect can be uncoupled from copy number-dependent expression in transgenic plants. *Plant Cell* **7**: 599–609.

Mlynarova L, Hricova A, Loonen A and Nap JP (2003) The presence of a chromatin boundary appears to shield a transgene in tobacco from RNA silencing. *Plant Cell* **15**: 2203–2217.

Ohlsson R, Renkawitz R and Lobanenkov V (2001) CTCF is a uniquely versatile transcription regulator linked to epigenetics and disease. *Trends Genet* **17**: 520–527.

Paul AL and Ferl RJ (1998) Higher order chromatin structures in maize and *Arabidopsis*. *Plant Cell* **10**: 1349–1359.

Peters AHFM, OCarroll D, Scherthan H *et al.* (2001) Loss of the Suv39h histone methyltransferases impairs mammalian heterochromatin and genome stability. *Cell* **107**: 323–337.

Phi-Van L, von Kries JP, Ostertag W and Stratling WH (1990) The chicken lysozyme 5 matrix attachment region increases transcription from a heterologous promoter in heterologous cells and dampens position effects on the expression of transfected genes. *Mol Cell Biol* **10**: 2302–2307.

Rea S, Eisenhaber F, O'Carroll D *et al.* (2000) Regulation of chromatin structure by site-specific histone H3 methyltransferases. *Nature* **406**: 593–599.

Recillas-Targa F, Pikaart MJ, Burgess-Beusse B, Bell AC, Litt MD, West AG, Gaszner M and Felsenfeld G (2002) Position-effect protection and enhancer blocking by the chicken beta-globin insulator are separable activities. *Proc Natl Acad Sci USA* **99**: 6883–6888.

Reeves R (2001) Molecular biology of HMGA proteins: hubs of nuclear function. *Gene* **277**: 63–81.

Romig H, Fackelmayer FO, Renz A, Ramsperger U and Richter A (1992) Characterization of SAF-A, a novel nuclear DNA binding protein from HeLa cells with high affinity for nuclear matrix/scaffold attachment DNA elements. *EMBO J* **11**: 3431–3440.

Scheuermann RH and Chen U (1989) A developmental-specific factor binds to suppressor sites flanking the immunoglobulin heavy-chain enhancer. *Genes Dev* **3**: 1255–1266.

Shübeler D, Mielke C, Maass K and Bode J (1996) Scaffold/matrix-attached regions act upon transcription in a context-dependent manner. *Biochemistry* **35**: 11160–11169.

Singh GB, Kramer JA and Krawetz SA (1997) Mathematical model to predict regions of chromatin attachment to the nuclear matrix. *Nucleic Acids Res* **25**: 1419–1425.

Tikhonov AP, Lavie L, Tatout C, Bennetzen JL, Avramova Z and Deragon JM (2001) Target sites for SINE integration in Brassica genomes display nuclear matrix binding activity. *Chromosome Res* **9**: 325–337.

Weitzel JM, Buhrmester H and Stratling WH (1997) Chicken MAR-binding protein

ARBP is homologous to rat methyl-CpG-binding protein MeCP2. *Mol Cell Biol* **17**: 5656–5666.

Whitehurs C, Henney HR, Max EE, Schroeder Jr HW, Stuber F, Siminovitch KA and Garrard WT (1992) Nucleotide sequence of the intron of the germline human kappa immunoglobulin gene connecting the J and C regions reveals a matrix association region (MAR) next to the enhancer. *Nucleic Acids Res* **20**: 4929–4930.

Zahn-Zabal M, Kobr M, Girod PA, Imhof M, Chatellard P, deJesus M, Wurm F and Mermod N (2001) Development of stable cell lines for production or regulated expression using matrix attachment regions. *J Biotechnol* **87**: 29–42.

Targeted gene insertion to enhance protein production from cell lines

3

Trevor N. Collingwood and Fyodor D. Urnov

3.1 Introduction

Critical to large-scale recombinant protein production is the generation and characterization of clonal cell lines that exhibit sustained, high level, expression of the recombinant gene of interest. High expression, as well as good cell growth and viability, are primary requirements for producer cell lines, yet generation of cell lines that fit these criteria represents a significant bottleneck in bioprocess. The >1 year timeline commonly experienced for isolation of a final producer clone represents a costly component of protein therapeutic development – at least in terms of time to market. In addition to the time cost are the physical resources required to screen hundreds or thousands of clones in order to identify those few that exhibit the desired growth and protein expression properties. This need for large-scale clonal screening is a consequence of the limitations of traditional methods used for generation of recombinant producer cell lines.

Historical approaches to generating high-producer cell lines have used random gene insertion into the recipient cell line. Typically, this was achieved by transfection of the cells with expression constructs containing both the gene of interest and a positive selection marker – each driven by a separate heterologous promoter. Subjecting the cells to marker selection would then result in the isolation of thousands of clones in which the marker gene and the gene of interest were stably co-expressed and subsequently co-replicated with cell division. A key facet of this approach is that the transgenes become stably integrated into random loci within the host cell's genome.

Random transgene integration has a major drawback with respect to generating high quality protein-producing cell lines: that is, the occurrence of positional effects of integration. The transgene can insert, theoretically, into any region of the host genome. However, the vast majority of the mammalian genome is estimated to exist in the form of densely compacted and transcriptionally refractory heterochromatin. Insertion into such a region is likely to result in the silencing of transgene expression due to encroaching repressive chromatin structure. While the successful isolation of selection-resistant clones implies viable transgene expression, it provides no guarantee as to the level or the temporal stability of expression, nor desirable growth characteristics of the cell line. For the majority of clones, expression of the transgene is either too low to be of value, or becomes silenced following removal of the selection pressure. Furthermore, false positive clones arise that are resistant to selection due to incorporation of the intact resistance marker, but in which the expression cassette of the gene of interest has segregated from the marker. It seems likely that silencing would be less of a concern if the transgene is inserted into loci that are highly transcriptionally active, such as those of housekeeping genes. This, however, presents its own problem in that the endogenous genes in such loci are generally necessary for maintaining cell growth and viability. Random transgene insertion into, or adjacent to, a critical gene may disrupt its expression and thereby have serious consequences for cell growth.

For the generation of any recombinant cell line the positional effects of integration severely limit the proportion of high quality clones. This necessitates the screening of large numbers of clones to identify those few that fulfill the desired functional criteria. To avoid these positional effects, transgenes must be inserted into the host cell genome in a way that does not significantly interfere with endogenous gene expression, or result in the shutdown of transgene expression by epigenetic mechanisms.

Several new methods exist which address some aspects of positional effects. Such approaches include mammalian artificial chromosomes (MACs) (Lindenbaum et al., 2004), scaffold-matrix attachment regions (Girod et al., 2005), and various genomic bordering elements (Burgess-Beusse et al., 2002; Kwaks and Otte 2006). These methods mostly provide protection of the integrated transgene from chromatin-dependent silencing, although the use of MACs also avoids disruption of essential endogenous genes. The methods for protection of the integrated transgene are described in Chapter 2.

This chapter, however, will focus solely on methods for the site-specific insertion of transgene expression cassettes precisely into defined, biologically innocuous loci that are relatively free from positional effects. Such genomic locations are referred to as 'hot spot' or 'safe harbor' loci. Site-specific transgene integration into a native endogenous locus can be achieved using homologous recombination (HR), whereby the gene to be delivered is flanked by a sequence that is homologous to the target locus. However, with the exception of certain cell types, such as mouse embryonic stem cells, the normal frequency of HR is in the order of one targeting event per 10^5 to 10^7 cells, and is 1000-fold lower than the frequency of random integration (Vasquez et al., 2001). Thus, for the purposes of recombinant protein production, HR has not been a viable approach to targeted integration in the context of protein production cell lines.

Targeted gene insertion requires two essential elements: (i) knowledge of the precise genomic location, or hot spot, into which the transgene is to be inserted; and (ii) a technology to effect targeted gene delivery. In this review we describe initially the methods by which hot spot loci can be identified, then follow with discussion of the current broadly used methods for targeted gene insertion. Finally, we extend our discussion to include emerging technologies that represent new approaches to targeted gene insertion and which could offer increased flexibility and efficiency of site-specific gene insertion, as well as potentially reducing the time cost for producer cell line generation.

3.2 Identification of genomic 'hot spot' loci

The identification of hot spots for transgene expression can be achieved by the random integration approach mentioned above. This requires the delivery of an expression cassette tagged with a quantifiable reporter gene and a selectable marker, then isolation and functional characterization of the stable clones. Clones that demonstrate the critical requirements of continuous high-level expression of the reporter and negligible effects on cell growth do, by definition, delineate a hot spot locus into which the expression cassette has integrated. As we discuss below, contemporary targeting approaches may then be used in a second step to insert a desired expression construct into the pre-integrated marker cassette.

Alternatively, for newer insertion technologies that target the native, unmodified locus directly (see below), it is a trivial task to clone and sequence the genomic DNA that flanks the inserted expression cassette and thus identify the precise genomic address of the hot spot. Subsequent single-step targeting with the gene of interest directly into the very same native locus would be expected to engender the same stability of expression and growth characteristics as seen with the marker gene, but without the need to first insert the tagging cassette.

The use of random integration to initially identify a hot spot might at first appear to defeat the purpose of single-step targeted integration. However, once a desired locus is identified it is then possible to repeatedly target that predetermined native locus with any new transgene. Furthermore, there is now an increasing number of protein biologics – either in development or in the commercial production – that are expressed in high-producer cells lines generated by random integration and selection. The expression construct in each of these cell lines already effectively marks a hot spot locus and, again, it would be possible to determine the exact genomic sequence and location of the insertion site so that they may be retargeted with other transgenes.

In either case, following hot spot identification, an efficient method is needed for targeting the insertion of the gene of interest into such sites.

3.3 Recombinase-mediated site-specific gene insertion

The most prevalent methods for targeted gene insertion in production cell lines focus on the use recombinases such as Cre, Flp, and ϕC31 that catalyze reciprocal site-specific recombination. By this approach, (i) integrated sequences that are flanked by recombinase sites can be excised from the genome; and (ii) in the reverse reaction, exogenous transgene expression cassettes that are flanked by specific recombinase sites can be inserted into similar sites pre-engineered into an endogenous locus. A relatively recent evolution of this mechanism involves the switching out of an integrated marker gene in conjunction with the simultaneous insertion of a recombinant gene of interest that is supplied by a donor targeting plasmid (Baer and

Bode, 2001; Coroadinha *et al.*, 2006; Schlake and Bode, 1994; Unsinger *et al.*, 2004; Wirth and Hauser, 2004). This is the process of recombination-mediated cassette exchange (RMCE) that has become a cornerstone technology for repeatedly targeting gene insertion into pre-defined genomic loci.

3.3.1 Cre, Flp, and φC31 recombinase systems

Cre and Flp recombinases have been the workhorses for genetic manipulation since their initial characterization over 20 years ago (Abremski *et al.*, 1983; Andrews *et al.*, 1985; Hoess and Abremski, 1984). Cre derives from the P1 bacteriophage, and FLP derives from *Saccharomyces cerevisiae*. Both belong to the λ integrase family of site-specific recombinases. Their mechanisms of action are very similar, with each catalyzing recombination between copies of its respective recognition element: loxP for Cre, and FRT (Flp Recombination Target) for Flp. In mammalian cells Cre is the more active recombinase, which is likely due to its greater thermal stability (Buchholz *et al.*, 1996). However, cycling mutagenesis has generated the FlpE mutant that exhibits enhanced thermal stability and function in mammalian cells (Buchholz *et al.*, 1998; Rodriguez *et al.*, 2000).

The canonical loxP and FRT elements are both comprised of two 13-bp inverted repeats flanking an asymmetric 8-bp core spacer that determines the directionality of the site, but the respective sequences differ between loxP and FRT. In addition, FRT contains an extra 13-bp direct repeat (Andrews *et al.*, 1985). The recombinases bind to each of the cognate repeats in a homodimeric manner and, during recombination events, they cleave the intervening spacer DNA at the margins to generate 8-bp 5' protruding ends (Andrews *et al.*, 1985). The overhangs are important determinants of the site-specificity of recombination with other similarly cleaved loxP or FRT sites present in either the same (*cis*) or separate (*trans*) DNA molecules.

The integrase from the *Streptomyces* actinophage φC31 provides a further avenue for recombinase-mediated gene insertion. The natural role of this enzyme is to drive integration of the φC31 phage into the *Streptomyces* genome through recombination between the attachment sites attP and attB in the phage and bacterial genomes, respectively (Rausch and Lehmann 1991). These sites are comprised of 3-bp core sequences flanked by inverted repeats that differ between attP and attB. The minimal attP and attB sites have been defined as 39-bp and 34-bp in size, respectively (Groth and Calos 2004; Groth *et al.*, 2000). In mammalian cells, φC31 mediates a process whereby transgenes flanked by attB sites can be efficiently integrated into attP sites that have been preinserted into endogenous chromosomes (Groth *et al.*, 2000; Thyagarajan *et al.*, 2001).

3.3.2 Recombinase-mediated cassette exchange

The most recent adaptation of the recombinase-based mechanisms described above is the process of recombination-mediated cassette exchange (RMCE) (Baer and Bode, 2001; Wirth and Hauser, 2004). The sophistication of RMCE approaches continues to develop and for this review we summarize and provide an example of current iterations of the method. From a practical standpoint, the central mechanism of RMCE

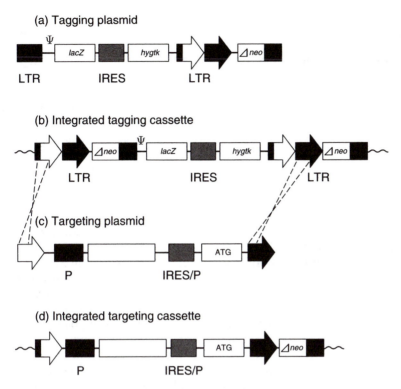

Figure 3.1

Tag and target methodology for cassette exchange by Flp mediated recombination. (A) Schematic representation of the tagging retroviral plasmid containing a lacZ reporter gene, hygtk positive/negative selection markers, and two FRT sites including one wild-type (white arrow) and one mutant (black arrow) located in the 3' LTR, followed by a defective neo gene. (B) Tagging construct after proviral integration, resulting in a duplication and transfer of the two FRT sites to the 5' LTR. (C) Schematic representation of the tagging plasmid containing the two FRT sequences flanking the gene of interest and an ATG sequence, which will restore the neo gene in the tagged clone after Flp recombinase mediated exchange (D). Reprinted from Coroadinha AS, et al. (2006) J Biotechnol 124: 457–468, with permission from Elsevier.

involves two key steps. A recently reported (Coroadinha *et al.*, 2006) example of RMCE is presented in *Figure 3.1*, and serves as a general template to this approach. The first step involves concomitant tagging of a hot spot locus with a marker gene, and insertion of recombinase target sites into the marked genomic locus. This is achieved by delivering a single construct containing both the marker gene and the recombinase binding sites. Construct delivery can be achieved using retrovirus, which can be titrated to increase the likelihood of single copy integration of the tagging vector (Wirth and Hauser, 2004), thus better ensuring consistency of expression for subsequently inserted transgenes. However, nonviral plasmid delivery is also an option, yet by either method isolation and characterization of the ensuing clonal cell lines is required prior to advancing to the next step. For the retroviral tagging vector, the recombinase target sites consist of two heterospecific sites (described below) for the specified recombinase located in the 3' long terminal repeat (LTR) of the vector (*Figure 3.1A*). In the present example, the heterospecific sites are for the FlpE recombinase and include FRT (wt), and the mutant FRT (F5) that contains mutations within the spacer sequence between the repeats (Schlake and Bode, 1994). The F5 mutant sites recognize each other with good specificity but show little cross-reaction with the wild-type FRT sites. Following viral transduction, the natural process of duplication of the regulatory U3 region within the LTR will result in a further copy of the heterospecific recombinase sites being placed in the 5' LTR upon retroviral replication (*Figure 3.1B*) (Baer and Bode, 2001). Within the boundaries of the 5' and 3' recombinase sites, the tagging insert contains a positive/ negative selectable marker gene, typically the hygromycin B phosphotransferase/thymidine kinase fusion protein (*hygtk*), and a quantifiable reporter, such as β-galactosidase (*lacZ*). Both genes can be driven off the LTR promoter if they are linked bicistronically by an internal ribosomal entry site (IRES) (Verhoeyen *et al.*, 2001). Thus, stable integrants first can be selected for by hygromycin, then the ability of each clone to support high level protein expression can be assessed by measuring β-galactosidase levels. This step achieves the combined goals of identifying hot spot loci and inserting a genomic landing pad (the recombinase sites) at those loci for subsequent transgene insertion.

In the second step, the transgene of interest is delivered by a targeting vector and is inserted into the recombinase sites in the clones selected above. In the targeting vector, the gene of interest is driven off a powerful heterologous promoter, such as that of cytomegalovirus (CMV) (*Figure 3.1C*). This expression construct is flanked by different single-specificity recombinase sites: in this example, a wild-type FRT (wt) site at one end, and a mutant FRT (F5) site at the other. In the presence of co-expressed FlpE recombinase (delivered on a plasmid), recombination occurs between identical sites, resulting in insertion of the target gene of interest. At this point, two simultaneous events occur: (i) excision from the integrated marker all of the sequences located between the recombinase sites; and (ii) insertion of all sequences between the captured recombinase sites in the targeting vector – including the gene of interest – into the integrated locus. The orientation of insertion is driven by the preference for FRT (wt) and FRT (F5) sites in the targeting vector to recombine with their respective exact matches at the integrated sites.

Selection for the integration event could be achieved by incorporating a constitutive selection marker within the integrating cassette. However, this does not present as a viable method for two reasons: (i) the high frequency of random integration events in protein production cell lines can lead to a significant level of false positives; and (ii) the RMCE insertion reaction is reversible in the continued presence of the recombinase, which can result in a further decrease in the ratio of successful targeted versus random integration events. The latter problem is to some extent mitigated by using a large excess of the targeting plasmid and thereby favoring insertion rather than excision. However, more advanced methods use a promoter trap approach that selects for only targeted integration events. Different experimental designs can achieve this goal, but in the example shown in *Figure 3.1*, the initial retroviral construct contains a promoterless neomycin resistance marker (Δneo) in the 3′ LTR (*Figure 3.1A*). Upon successful site-specific recombination with the targeting cassette, the Δneo gene becomes functionally linked by an IRES and an initiation codon to the same promoter that drives expression of the gene of interest (*Figure 3.1D*). This then activates expression of the *neo* marker and allows selection for correct RMCE events using G418. Since the excision event from integrated viral construct removes the *hygtk* marker, any remaining cells in which RMCE has not occurred can be removed by negative selection via activity of the thymidine kinase (*tk*) gene in the presence of gancyclovir. The efficiency of RMCE when using such a selection strategy is high, with generally >90% of the G418-resistant clones exhibiting correct insertion events (Wirth and Hauser, 2004).

3.3.3 Gene insertion at native 'pseudo' recombinase sites

Probability dictates that the 34–39-bp minimal loxP/FRT and attB/attP sites do not occur naturally in the mammalian genome (Nagy, 2000) and must therefore be pre-engineered into the target genome. However, endogenous 'pseudo' lox and att sites have been identified in the mouse and human genomes that have partial sequence identity with their wild-type elements and support recombinase-mediated integration – albeit with reduced efficiency compared with the wild-type sequences (Thyagarajan *et al.*, 2000, 2001). A directed evolution approach using DNA shuffling has been able to generate φC31 mutants with enhanced activity on specific pseudo attP sites (Sclimenti *et al.*, 2001). More recently, a study of φC31-mediated integration into pseudo-attP sites in human cell lines found that of 196 independent integration events, 101 different integration sites were used, with the majority being distributed among 19 pseudo attP sites (Chalberg *et al.*, 2005). This group of pseudo attP sites revealed a ~30-bp palindromic consensus sequence, supporting some degree of sequence specificity for these φC31-mediated targeting events.

The presence of naturally occurring pseudo recombination sites may provide an opportunity to use recombinases to target gene insertion directly into native genomic locations in a single step, without the need for prior binding site insertion. This could be useful in the field of protein production if such sites could be shown to reside within genomic hot spots. However, the apparent promiscuity of these prokaryotic recombinases in the presence of a mammalian-sized genome could also give rise to unintended insertion or translocation events. It would thus be important to characterize the cell lines to ensure that no events have occurred that are detrimental to protein production.

3.3.4 Modification of recombinases and their target sites

Gene insertion into single wild-type recombinase sites faces the problem of subsequent re-excision due to the reverse reaction in the continued presence of the transiently expressed recombinase (Baer and Bode, 2001). The excision event is driven by intragenic recombination between the still-present recombinase sites that flank the inserted transgene, thereby removing the newly recombined gene. The use of heterospecific recombinase sites, in which the 5' and 3' sites differ, has been invaluable in overcoming this problem. In addition to the mutant FRT sites mentioned in the example above (Coroadinha *et al.*, 2006; Schlake and Bode, 1994), further studies with mutant loxP/FRT elements are providing alternative solutions to the problem of transgene excision. The introduction of nucleotide changes into either the left (LE mutant, lox71) or right (RE mutant, lox66) 13-bp inverted repeats in loxP results in asymmetric Cre recognition sites that are still cleavable by the recombinase (Albert *et al.*, 1995). Recombination between LE or RE mutant sites in the recipient genomic locus and RE or LE mutant sites in the targeting construct generates LE+RE double mutant lox sites. The composite mutant sites are poorly recognized by Cre. This greatly reduces the Cre-mediated reverse excision reaction, enabling essentially irreversible and stable RMCE into the mutant lox sites. However, the efficiency of recombination between homologous mutant sites was also considerably reduced. While initial studies were performed in plants, this approach has since been extended to mammalian cell systems (Araki *et al.*, 1997). Others have targeted mutation of the 8-bp spacer in the loxP site to generate asymmetry and thereby reduce the frequency of transgene excision (Hoess *et al.*, 1986; Lee and Saito, 1998). Mutagenesis-based strategies continue to identify new variants of the loxP/FRT sites (Thomson *et al.*, 2003), while the generation of Cre and Flp variants with altered recognition site specificity is also providing increased flexibility of targeting with these systems (Buchholz *et al.*, 1996; Jayaram *et al.*, 2004; Konieczka *et al.*, 2004; Saraf-Levy *et al.*, 2006; Voziyanov *et al.*, 2003).

A major mechanistic difference, however, between wild-type φC31 and the Cre and Flp recombinases is that φC31-mediated transgene integration is unidirectional. This is because recombination between attP and attB sites leads to generation of hybrid attL and attR sites that flank the newly integrated cassette and which are no longer recognized by the φC31 integrase (Thorpe and Smith, 1998). The absence of a competing excision reaction thereby favors stable site-specific φC31-directed integration. A comparative study of the relative ability of wild-type φC31, Cre, and Flp recombinases to cleave integrated substrates and promote intramolecular recombination in mammalian cells revealed Cre/LoxP to be the more powerful system. It was suggested that this may be due to the presence of an intrinsic bipartite nuclear localization sequence that facilitates transport of this prokaryotic protein to the mammalian nucleus (Le *et al.*, 1999). The subsequent addition of a nuclear localization sequence to the carboxy terminus of φC31 integrase increased its relative activity from ~10% to >50% that of Cre recombinase, and to a level considerably higher than that of the FlpE mutant (Andreas *et al.*, 2002).

It is clear that the two-step recombinase-based methods for targeted gene integration continue to evolve. However, the advent of single-step integration technologies such as the engineered zinc finger nuclease approach described below is expected to significantly augment the genetic tool selection available to achieve site-specific gene insertion.

3.4 Emerging technologies for targeted gene insertion

In eukaryotic cells, the naturally occurring process of eliminating double-strand DNA breaks (DSBs) (Hoeijmakers, 2001) provides another solution to the problem of site-specific gene insertion. The DSB is one of the most common types of DNA damage that affect eukaryotic cells. It arises due to replication fork collapse in S phase at a rate of 5–50 times/cell/division cycle, or environmental insult, such as X-rays (Valerie and Povirk, 2003). To protect against this genetic damage, all eukaryotes use highly conserved and efficient pathways for DSB repair which are rapidly evoked after the break is detected. This process has been shown to be exceptionally effective and specific in the context of two distinct strategies for targeted transgene insertion: (i) use of homing endonucleases (Belfort and Roberts, 1997) on chromosomal targets that have been previously engineered to contain cognate recognition sites for such endonucleases; and (ii) use of designed zinc finger nucleases (Bibikova *et al.*, 2003) on native, intact chromosomal loci chosen by the investigator. Importantly, aside from the difference between the two approaches in terms of the requirement for chromosome engineering, both methods invoke the same core intranuclear process – that of DSB-induced homology-directed repair (HDR) (Symington, 2002).

In its usual context, HDR involves the exact restoration of DNA sequence lost at the break by copying genetic information from a homologous DNA sequence – usually the sister chromatid. This natural repair pathway can be directly exploited to effect high-frequency site-specific transgene insertion into a previously unmodified genomic locus. Following a DSB, HDR begins with resection of the two broken DNA ends to yield single-stranded overhangs on either side of the break (*Figure 3.2*). The gap thereby generated in the genome is repaired as follows: the single-stranded tails are uniformly coated with Rad51 to yield a 'presynaptic fiber.' This molecular entity then performs a 'homology search' for an intact double-stranded DNA molecule carrying an identical sequence. This is the point at which the DNA repair process can be usurped to effect site-specific insertion of an exogenous transgene expression cassette. Instead of allowing the cell to use the sister chromatid as a repair template, the cell can be flooded with an exogenous donor molecule that contains the transgene. The transgene is flanked by DNA that is homologous to the sequence which flanks the DSB. The transgene-containing donor DNA molecule is then used as a template to repair the broken chromosome by HDR. While the frequency of homologous recombination is normally in the order of 10^{-6}, it is increased by two to four orders of magnitude during HDR (Symington, 2002). The net result of this repair process is the integration of the entire transgene into the chromosome precisely at the location of the DSB. The HDR process appears to be affected little by the presence of nonhomologous sequence (e.g. a transgene) between the two homologous ends (Nassif *et al.*, 1994).

A fundamental mechanistic distinction exists between the unidirectional transfer of DNA sequence information observed during HDR and conventional homologous recombination, i.e. during meiosis. The latter, while also initiated by DSB, leads to an exchange of chromosome arms and involves the resolution of two Holliday junctions. DSB repair in somatic cells, however, does not generate a classical crossover event: the broken and unbroken chromosomes remain intact, with the exception of genetic information that is unidirectionally transferred from one to the other (Richardson *et al.*, 1998). This chromosomal stability during DSB-mediated HDR is an important consideration when applying the technology to transgene insertion.

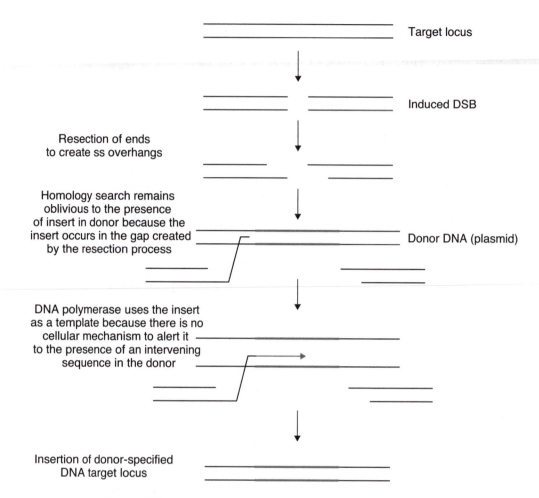

Target locus

Induced DSB

Resection of ends
to create ss overhangs

Homology search remains
oblivious to the presence
of insert in donor because the
insert occurs in the gap created
by the resection process

Donor DNA (plasmid)

DNA polymerase uses the insert
as a template because there is no
cellular mechanism to alert it
to the presence of an intervening
sequence in the donor

Insertion of donor-specified
DNA target locus

Figure 3.2

Double-stranded DNA break (DSB)-induced targeted integration. An induced
DSB is followed by 3′ to 5′ resection of the cleaved DNA ends then a homology
search for donor DNA. A synthesis-dependent strand annealing mechanism
transfers the donor DNA molecule (including its transgene payload) into the
chromosome (Nassif *et al.*, 1994).

A critical limitation to gene insertion by the HDR-mediated process is the
requirement for a DSB to be generated at the precise location into which the
gene is to be inserted. Hence, a method for generating targeted DSBs at
endogenous loci is a primary requisite for this approach. The use of site-
specific nucleases described below provides an approach by which this goal
can be achieved.

3.4.1 Homing endonucleases in HDR-mediated targeted gene insertion

The mitochondrial genome of yeast encodes a homing endonuclease, I-SceI,
which cleaves an 18-bp DNA sequence to create a DSB. The broken ends

home in to either side of the open reading frame encoding the nuclease itself, which then relocates to the position of the DSB (Dujon *et al.*, 1986; Monteilhet *et al.*, 1990). I-SceI and its many relatives are highly sequence-specific nucleases that are well tolerated by most cells because mammalian genomes do not contain naturally occurring recognition sites for those enzymes. For this reason, a two-step procedure (*Figure 3.3*) has been developed to apply the homing endonucleases to genome manipulation (Rouet *et al.*, 1994).

A target chromosome is first engineered (typically by random integration) to contain the recognition sequence for a homing endonuclease. These genetically engineered cells are then used as substrates for a second step in which the cells are exposed to an expression construct encoding the I-SceI endonuclease itself and a 'donor' DNA construct. The latter molecule will typically contain a region of homology to the engineered chromosome that is centered on the I-SceI site (*Figure 3.3*), but into which the transgene payload has been inserted. Experiments from a large number of laboratories – and importantly, in multiple species – have consistently shown that an I-SceI-induced DSB in the chromosome potentiates the rate of transgene integration into the chromosomal location by two to three orders of magnitude (Johnson and Jasin, 2001). Insertion of the transgene targeting construct into the cells can be followed by drug-based selection to identify the clones in which the desired homologous recombination event has occurred.

A significant limitation to I-SceI-mediated targeted integration method in somatic cells is that, as with the Cre, Flp, and φC31 recombinases discussed earlier, it requires its recognition site to be first stably inserted into the target locus by random integration methods. The desire to overcome the requirement for this first step has prompted an effort to re-engineer these enzymes to recognize DNA sequences that occur at native endogenous genomic locations (Arnould *et al.*, 2006; Chames *et al.*, 2005; Chen and Zhao 2005; Epinat *et al.*, 2003). It will be of interest to see how amenable the homing endonucleases are to such engineering.

3.4.2 Targeted gene insertion into native loci by zinc finger nuclease-mediated, high-frequency, homologous recombination

The challenge of efficiently targeting transgene integration into a native locus appears resolved by a new advance in the homology-directed repair approach described earlier. The novel feature here is the ability to generate a DSB at essentially any specific site within any mammalian genome. Once the DSB is generated, HDR can effect transgene insertion by high frequency homologous recombination. The capacity to induce site-specific DSB is due to the development of engineered zinc finger nucleases (ZFNs; also known in the literature as 'chimeric endonucleases'). As described in more detail below, a catalytically active nuclease domain is fused to an engineered site-specific DNA binding domain that directs the nuclease to the desired genomic address. The nuclease then generates the site-specific DSB, and HDR follows.

The two critical components of a ZFN are the DNA binding domain and the catalytic nuclease domain. The former makes use of the most common

DNA binding domain in metazoan genomes – the Cys_2-His_2 zinc finger (Klug, 1995; Miller *et al.*, 1985). This protein motif has two properties that make it useful for directing site-specific genome modification. First, the protein–DNA interface is relatively simple and well-understood (Pavletich

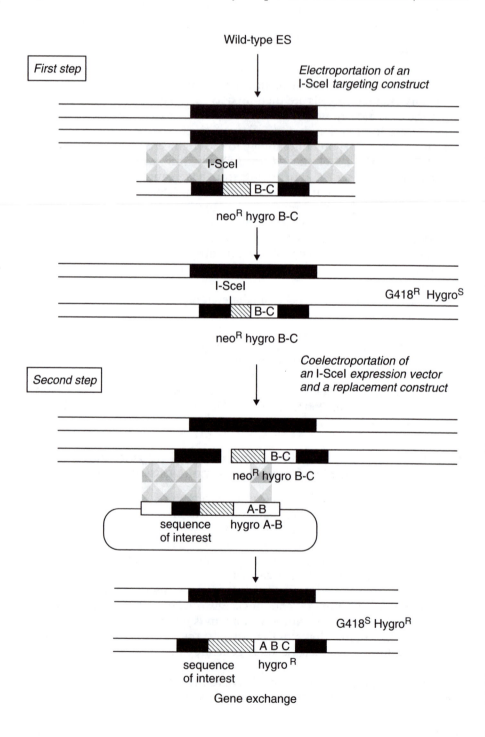

and Pabo, 1991). Each zinc finger recognizes 3–4 bp of DNA. Several fingers (typically 3–6) can be linked together to generate a recognition module with composite specificity for 9–18-bp sequences, respectively (Isalan and Choo, 2001). A large collection of designed zinc fingers has been generated that can be used to create a DNA binding module to target essentially any sequence (Pabo *et al.*, 2001). A second important property of these proteins is that a tandem array of zinc fingers behaves as an integral DNA-binding unit; that is, it effectively binds DNA even when fused to heterologous proteins, including enzyme catalytic domains (Kadonaga *et al.*, 1987, 1988; Reik *et al.*, 2002; Snowden *et al.*, 2002, 2003).

The catalytic activity of the ZFN results from fusing the zinc finger protein DNA binding domain to the cleavage domain of the type IIS endonuclease, FokI, to yield a zinc finger nuclease with the ability to induce a double-strand break at a DNA sequence specified by the zinc finger module (Kim *et al.*, 1996). As with most restriction enzymes, FokI binds to its target site as a homodimer, with each catalytic domain breaking one strand of the DNA. This is important in the context of ZFNs described here because DSB formation requires not one, but two distinct zinc finger nucleases to bind to the prescribed DNA stretch – one on either side of the position of the desired DSB, and with the appropriate spacing (*Figure 3.4*). This stringent requirement for a precise juxtaposition of the two ZFNs in order to effect the DSB reduces considerably the possibility for off-target DSB events. This is because the catalytic activity of the FokI domain first requires DNA binding of adjacent ZFNs to allow formation of a FokI domain dimer with DNA cleavage activity.

Seminal work has demonstrated that such ZFNs can induce a DSB in frog, insect, and mammalian cell chromosomes, and dramatically increase the frequency of repair of a disabled chromosomal reporter gene in whole animals using donor DNA molecules as templates (Bibikova *et al.*, 2001, 2002, 2003; Porteus and Baltimore, 2003). More recent studies focusing on the application of ZFNs to native loci in mammalian cells have demonstrated a targeted single base change in approximately 20% of the alleles of a target gene (Urnov *et al.*, 2005). Subsequent work in the same laboratory on more than 10 distinct endogenous loci supports the generality of this approach to genome engineering.

Figure 3.3 (opposite)

Two-step DSB-mediated targeted integration procedure as described by Cohen-Tannoudji et al. (1998). In the first step, the locus one wishes to drive integration into is modified, via conventional gene targeting, with the integration of an I-SceI (or other meganuclease) recognition site. Once a cell clone carrying the desired integrant has been identified, the second step involves a distinct targeting construct co-transfected into the cell along with an expression cassette for the nuclease. Selection for the correct integration event was possible due to restoration of a functional hygromycin resistance gene. Geometric shading denotes sequences of homology that drive the homologous recombination. (Copyright 1998 American Society for Microbiology. All rights reserved. Used by permission.)

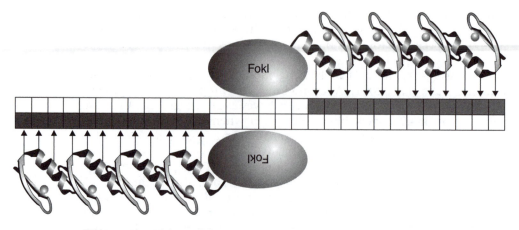

Figure 3.4

Schematic showing a ZFN dimer bound to a target locus to achieve a site-specific DSB. Each ZFN monomer shown contains four zinc fingers that are engineered to recognize a specific 12 base pair target sequence. To effect cleavage, each ZFN must bind to its cognate DNA target sequence, and precise spacing must exist between the bound moieties so that the catalytically active dimer can form.

In contrast to the single base changes described above, in the context of recombinant protein production the measure of success for a targeted integration technology is the capacity to insert an entire expression cassette for the gene of interest into a specific locus. Such expression cassettes are likely to be several kilobases in size. In studies aimed at ZFN-mediated integration of an 8-kb expression construct carrying three distinct promoter transcription units, we found that ~5% of the chromatids had the cassette integrated into the chromosomal location specified by the ZFNs. In the field of protein production, such a high frequency of targeted integration into a native locus should mean that site-specific integrants could be easily identified by screening a modest number (<100) of clones directly, even without selection. It is important to note that ZFN-mediated gene integration into a specified ZFN target site is irreversible, as it is with the use of I-SceI. This is because the dimeric nuclease binding site at the target locus is destroyed upon insertion of the transgene between the two monomer half-sites. The lack of a reverse excision reaction is no doubt an important element in the high overall efficiency of this approach.

3.5 Perspective

Site-specific transgene insertion into genomic hot-spot loci continues to represent an important avenue for reducing both the scale and the time cost of clonal screening for the isolation of high producer cell lines. New methods are evolving to increase both the efficiency and specificity of targeted integration, as well as providing greater flexibility in regard to choosing the site of integration. The capacity to target gene insertion effectively to any genomic address will become increasingly valuable as genetic studies teach us more about the location and character of natural endogenous hot-spot loci in different cell types and in different genomes.

The two-step RMCE approach to gene insertion described in this review is widely used and has been well validated. It is at present too early to assess the broad efficacy and utility of HDR-based approaches such as those represented by the homing endonucleases or designed zinc finger nucleases. However, early results are encouraging and the field is poised for success in this regard.

Parental cell lines such as those commonly used for protein production exhibit a significant degree of karyotypic heterogeneity. This is true not only for different isolates from a common mammalian species, but also within an isolated 'clonal' population – especially cells that have been immortalized or transformed (Derouazi et al., 2006; Fabarius et al., 2003; Namba et al., 1996). Two-step targeting approaches to gene insertion first involve selecting a subset of clonal isolates containing a marker gene that is stably, yet randomly, integrated at a locus into which the recombinant gene of interest can later be transferred. This initial selection, in effect, reduces the genetic – and perhaps even epigenetic – diversity of the cell population that will ultimately receive the gene of interest. The importance of the resulting loss of genetic diversity in terms of the range of expression phenotypes available for characterization is unclear. However, targeting gene insertion directly to nonmodified hot spot loci of the complete parental cell population – rather than a subset – should better preserve genetic diversity for an integration event at any given locus. It will be of interest to see what effect this may have on the range of expression characteristics exhibited by subsequently selected clones in which all isolates have the transgene targeted to the same locus, yet which may be otherwise genetically distinct. It is conceivable that this approach may provide the opportunity to identify clones in which the gene of interest resides in a more optimal genetic setting.

Although this review has focused on targeted integration of recombinant protein expression cassettes, it is worth noting that such site-directed genomic editing at native loci can also be used to effect other genetic outcomes, such as targeted gene knockout or gene modification. Gene knockout could be achieved either by the targeted insertion of disruptive sequences into the open reading frame of a gene, or by deletion of key portions of the target gene. Indeed, ZFN-mediated targeted disruption of an endogenous locus has recently been reported (Urnov et al., 2005). Alternatively, more subtle alteration of native gene sequences to generate isoforms with enhanced biological activity would also be possible using HDR-mediated processes. Together, such approaches provide the capacity not only to effect targeted insertion of transgene expression cassettes, but also to re-engineer producer cell lines to optimize other important characteristics such as cell growth and post-translation processing of the recombinant protein. Each of these parameters can provide important contributions to the yield and quality of the product and therefore the efficiency of the production process.

In addition to the highly reproducible observation of the rapid and efficient ZFN-driven targeted integration of an 8 kb expression cassette into a native locus in human cells, we now consistently find that a significant fraction of the integration events occur biallelically (Moehle et al., 2007).

References

Abremski K, Hoess R and Sternberg N (1983) Studies on the properties of P1 site-specific recombination: evidence for topologically unlinked products following recombination. *Cell* **32**: 1301–1311.

Albert H, Dale EC, Lee E and Ow DW (1995) Site-specific integration of DNA into wild-type and mutant lox sites placed in the plant genome. *Plant J* **7**: 649–659.

Andreas S, Schwenk F, Kuter-Luks B, Faust N and Kuhn R (2002) Enhanced efficiency through nuclear localization signal fusion on phage PhiC31-integrase: activity comparison with Cre and FLPe recombinase in mammalian cells. *Nucleic Acids Res* **30**: 2299–2306.

Andrews BJ, Proteau GA, Beatty LG and Sadowski PD (1985) The FLP recombinase of the 2 micron circle DNA of yeast: interaction with its target sequences. *Cell* **40**: 795–803.

Araki K, Araki M and Yamamura K (1997) Targeted integration of DNA using mutant lox sites in embryonic stem cells. *Nucleic Acids Res* **25**: 868–872.

Arnould S, Chames P, Perez C *et al.* (2006) Engineering of large numbers of highly specific homing endonucleases that induce recombination on novel DNA targets. *J Mol Biol* **355**: 443–458.

Baer A and Bode J (2001) Coping with kinetic and thermodynamic barriers: RMCE, an efficient strategy for the targeted integration of transgenes. *Curr Opin Biotechnol* **12**: 473–480.

Belfort M and Roberts RJ (1997) Homing endonucleases: keeping the house in order. *Nucleic Acids Res* **25**: 3379–3388.

Bibikova M, Carroll D, Segal DJ, Trautman JK, Smith J, Kim YG and Chandrasegaran S (2001) Stimulation of homologous recombination through targeted cleavage by chimeric nucleases. *Mol Cell Biol* **21**: 289–297.

Bibikova M, Golic M, Golic KG and Carroll D (2002) Targeted chromosomal cleavage and mutagenesis in Drosophila using zinc-finger nucleases. *Genetics* **161**: 1169–1175.

Bibikova M, Beumer K, Trautman JK and Carroll D (2003) Enhancing gene targeting with designed zinc finger nucleases. *Science* **300**: 764.

Buchholz F, Ringrose L, Angrand PO, Rossi F and Stewart AF (1996) Different thermostabilities of FLP and Cre recombinases: implications for applied site-specific recombination. *Nucleic Acids Res* **24**: 4256–4262.

Buchholz F, Angrand PO and Stewart AF (1998) Improved properties of FLP recombinase evolved by cycling mutagenesis. *Nat Biotechnol* **16**: 657–662.

Burgess-Beusse B, Farrell C, Gaszner M, Litt M, Mutskov V, Recillas-Targa F, Simpson M, West A and Felsenfeld G (2002) The insulation of genes from external enhancers and silencing chromatin. *Proc Natl Acad Sci USA* **99(Suppl 4)**: 16433–16437.

Chalberg TW, Portlock JL, Olivarer EC, Thyagarajan B, Kirby PJ, Hillman RT, Hoelters J and Calos MP (2006) Integration specificity of phage varphiC31 integrase in the human genome. *J Mol Biol* **357**: 28–48.

Chames P, Epinat JC, Guillier S, Patin A, Lacroix E and Paques F (2005) In vivo selection of engineered homing endonucleases using double-strand break induced homologous recombination. *Nucleic Acids Res* **33**: e178.

Chen Z and Zhao H (2005) A highly sensitive selection method for directed evolution of homing endonucleases. *Nucleic Acids Res* **33**: e154.

Cohen-Tannoudji M, Robine S, Choulika A, Pinto D, El Marjou F, Babinet C, Louvard D and Jaisser F (1998) I-SceI-induced gene replacement at a natural locus in embryonic stem cells. *Mol Cell Biol* **18**: 1444–1448.

Coroadinha AS, Schucht R, Gama-Norton L, Wirth D, Hauser H and Carrondo MJ (2006) The use of recombinase mediated cassette exchange in retroviral vector producer cell lines: Predictability and efficiency by transgene exchange. *J Biotechnol* **124**:457–468

Derouazi M, Martinet D, Besuchet Schmutz N, Flaction R, Wicht M, Bertschinger M, Hacker DL, Beckmann JS and Wurm FM (2006) Genetic characterization of CHO production host DG44 and derivative recombinant cell lines. *Biochem Biophys Res Commun* **340**: 1069–1077.

Dujon B, Colleaux L, Jacquier A, Michel F and Monteilhet C (1986) Mitochondrial introns as mobile genetic elements: the role of intron-encoded proteins. *Basic Life Sci* **40**: 5–27.

Epinat JC, Arnould S, Chames P, Rochaix P, Desfontaines D, Puzin C, Patin A, Zanghellini A, Paques F and Lacroix E (2003) A novel engineered meganuclease induces homologous recombination in yeast and mammalian cells. *Nucleic Acids Res* **31**: 2952–2962.

Fabarius A, Hehlmann R and Duesberg PH (2003) Instability of chromosome structure in cancer cells increases exponentially with degrees of aneuploidy. *Cancer Genet Cytogenet* **143**: 59–72.

Girod PA, Zahn-Zabal M and Mermod N (2005) Use of the chicken lysozyme 5' matrix attachment region to generate high producer CHO cell lines. *Biotechnol Bioeng* **91**: 1–11.

Groth AC and Calos MP (2004) Phage integrases: biology and applications. *J Mol Biol* **335**: 667–678.

Groth AC, Olivares EC, Thyagarajan B and Calos MP (2000) A phage integrase directs efficient site-specific integration in human cells. *Proc Natl Acad Sci USA* **97**: 5995–6000.

Hoeijmakers JH (2001) Genome maintenance mechanisms for preventing cancer. *Nature* **411**: 366–374.

Hoess RH and Abremski K (1984) Interaction of the bacteriophage P1 recombinase Cre with the recombining site loxP. *Proc Natl Acad Sci USA* **81**: 1026–1029.

Hoess RH, Wierzbicki A and Abremski K (1986) The role of the loxP spacer region in P1 site-specific recombination. *Nucleic Acids Res* **14**: 2287–2300.

Isalan M and Choo Y (2001) Rapid, high-throughput engineering of sequence-specific zinc finger DNA-binding proteins. *Methods Enzymol* **340**: 593–609.

Jayaram M, Mehta S, Uzri D, Voziyanov Y and Velmurugan S (2004) Site-specific recombination and partitioning systems in the stable high copy propagation of the 2-micron yeast plasmid. *Prog Nucleic Acid Res Mol Biol* **77**: 127–172.

Johnson RD and Jasin M (2001) Double-strand-break-induced homologous recombination in mammalian cells. *Biochem Soc Trans* **29(Pt 2)**: 196–201.

Kadonaga JT, Carner KR, Masiarz FR and Tjian R (1987) Isolation of cDNA encoding transcription factor Sp1 and functional analysis of the DNA binding domain. *Cell* **51**: 1079–1090.

Kadonaga JT, Courey AJ, Ladika J and Tjian R (1988) Distinct regions of Sp1 modulate DNA binding and transcriptional activation. *Science* **242**: 1566–1570.

Kim CA and Berg JM (1996) A 2.2 A resolution crystal structure of a designed zinc finger protein bound to DNA. *Nat Struct Biol* **3**: 940–945.

Klug A (1995) Gene regulatory proteins and their interaction with DNA. *Ann NY Acad Sci* **758**: 143–160.

Konieczka JH, Paek A, Jayaram M and Voziyanov Y (2004) Recombination of hybrid target sites by binary combinations of Flp variants: mutations that foster inter-protomer collaboration and enlarge substrate tolerance. *J Mol Biol* **339**: 365–378.

Kwaks TH and Otte AP (2006) Employing epigenetics to augment the expression of therapeutic proteins in mammalian cells. *Trends Biotechnol* **24**: 137–142.

Le Y, Gagneten S, Tombaccini D, Bethke B and Sauer B (1999) Nuclear targeting determinants of the phage P1 cre DNA recombinase. *Nucleic Acids Res* **27**: 4703–4709.

Lee G and Saito I (1998) Role of nucleotide sequences of loxP spacer region in Cre-mediated recombination. *Gene* **216**: 55–65.

Lindenbaum M, Perkins E, Csonka E, Fleming E, Garcia L, Greene A, Gung L, Hadlaczky G, Lee E, Leung J *et al*. (2004) A mammalian artificial chromosome engineering system (ACE System) applicable to biopharmaceutical protein production, transgenesis and gene-based cell therapy. *Nucleic Acids Res* **32**: e172.

Miller J, McLachlan AD And Klug A (1985) Repetitive zinc-binding domains in the protein transcription factor IIIA from Xenopus oocytes. *EMBO J* **4**: 1609–1614.

Moehle EA, Rock JM, Lee YL, Jouvenot Y, DeKelver RC, Gregory PD, Urnov FD, and Holmes MC. Targeted gene addition: efficient insertion of a gene-sized DNA sequence into a specified location in the human genome using designed zinc finger nucleases. (*Submitted for publication*).

Monteilhet C, Perrin A, Thierry A, Colleaux L and Dujon B (1990) Purification and characterization of the in vitro activity of I-Sce I, a novel and highly specific endonuclease encoded by a group I intron. *Nucleic Acids Res* **18**: 1407–1413.

Nagy A (2000) Cre recombinase: the universal reagent for genome tailoring. *Genesis* **26**: 99–109.

Namba M, Mihara K and Fushimi K (1996) Immortalization of human cells and its mechanisms. *Crit Rev Oncog* **7**: 19–31.

Nassif N, Penney J, Pal S, Engels WR and Gloor GB (1994) Efficient copying of nonhomologous sequences from ectopic sites via P-element-induced gap repair. *Mol Cell Biol* **14**: 1613–1625.

Pabo CO, Peisach E and Grant RA (2001) Design and selection of novel Cys2his2 zinc finger proteins. *Annu Rev Biochem* **70**: 313–340.

Pavletich NP and Pabo CO (1991) Zinc finger-DNA recognition: crystal structure of a Zif268-DNA complex at 2.1 A. *Science* **252**: 809–817.

Porteus MH and Baltimore D (2003) Chimeric nucleases stimulate gene targeting in human cells. *Science* **300**: 763.

Rausch H and Lehmann M (1991) Structural analysis of the actinophage phi C31 attachment site. *Nucleic Acids Res* **19**: 5187–5189.

Reik A, Gregory PD and Urnov FD (2002) Biotechnologies and therapeutics: chromatin as a target. *Curr Opin Genet Dev* **12**: 233–242.

Richardson C, Moynahan ME and Jasin M (1998) Double-strand break repair by interchromosomal recombination: suppression of chromosomal translocations. *Genes Dev* **12**: 3831–3842.

Rodriguez CI, Buchholz F, Galloway J, Sequerra R, Kasper J, Ayala R, Stewart AF and Dymecki SM (2000) High-efficiency deleter mice show that FLPe is an alternative to Cre-loxP. *Nat Genet* **25**: 139–140.

Rouet P, Smith F and Jasin M (1994) Introduction of double-strand breaks into the genome of mouse cells by expression of a rare-cutting endonuclease. *Mol Cell Biol* **14**: 8096–8106.

Saraf-Levy T, Santoro SW, Volpin H, Kushnirsky T, Eyal Y, Schultz PG, Gidoni D and Carmi N (2006) Site-specific recombination of asymmetric lox sites mediated by a heterotetrameric Cre recombinase complex. *Bioorg Med Chem* **14**:3081–3089.

Schlake T and Bode J (1994) Use of mutated FLP recognition target (FRT) sites for the exchange of expression cassettes at defined chromosomal loci. *Biochemistry* **33**: 12746–12751.

Sclimenti CR, Thyagarajan B and Calos MP (2001) Directed evolution of a recombinase for improved genomic integration at a native human sequence. *Nucleic Acids Res* **29**: 5044–5051.

Snowden AW, Gregory PD, Case CC and Pabo CO (2002) Gene-specific targeting of H3K9 methylation is sufficient for initiating repression *in vivo*. *Curr Biol* **12**: 2159–2166.

Snowden AW, Zhang L, Urnov F, Dent C, Jouvenot Y, Zhong X, Rebar EJ, Jamieson AC, Zhang HS, Tan S *et al*. (2003) Repression of vascular endothelial growth factor A in glioblastoma cells using engineered zinc finger transcription factors. *Cancer Res* **63**: 8968–8976.

Symington LS (2002) Role of RAD52 epistasis group genes in homologous recombination and double-strand break repair [table of contents]. *Microbiol Mol Biol Rev* **66**: 630–670.

Thomson JG, Rucker 3rd EB and Piedrahita JA (2003) Mutational analysis of loxP sites for efficient Cre-mediated insertion into genomic DNA. *Genesis* **36**: 162–167.

Thorpe HM and Smith MC (1998) In vitro site-specific integration of bacteriophage DNA catalyzed by a recombinase of the resolvase/invertase family. *Proc Natl Acad Sci USA* **95**: 5505–5510.

Thyagarajan B, Guimaraes MJ, Groth AC and Calos MP (2000) Mammalian genomes contain active recombinase recognition sites. *Gene* **244**: 47–54.

Thyagarajan B, Olivares EC, Hollis RP, Ginsburg DS and Calos MP (2001) Site-specific genomic integration in mammalian cells mediated by phage phiC31 integrase. *Mol Cell Biol* **21**: 3926–3934.

Unsinger J, Lindenmaier W, May T, Hauser H and Wirth D (2004) Stable and strictly controlled expression of LTR-flanked autoregulated expression cassettes upon adenoviral transfer. *Biochem Biophys Res Commun* **319**: 879–887.

Urnov FD, Miller JC, Lee YL, Beausejour CM, Rock JM, Augustus S, Jamieson AC, Porteus MH, Gregory PD and Holmes MC (2005) Highly efficient endogenous human gene correction using designed zinc-finger nucleases. *Nature* **435**: 646–651.

Valerie K and Povirk LF (2003) Regulation and mechanisms of mammalian double-strand break repair. *Oncogene* **22**: 5792–5812.

Vasquez KM, Marburger K, Intody Z and Wilson JH (2001) Manipulating the mammalian genome by homologous recombination. *Proc Natl Acad Sci USA* **98**: 8403–8410.

Verhoeyen E, Hauser H and Wirth D (2001) Evaluation of retroviral vector design in defined chromosomal loci by Flp-mediated cassette replacement. *Hum Gene Ther* **12**: 933–944.

Voziyanov Y, Konieczka JH, Stewart AF and Jayaram M (2003) Stepwise manipulation of DNA specificity in Flp recombinase: progressively adapting Flp to individual and combinatorial mutations in its target site. *J Mol Biol* **326**: 65–76.

Wirth D and Hauser H (2004) Flp-mediated integration of expression cassettes into FRT-tagged chromosomal loci in mammalian cells. *Methods Mol Biol* **267**: 467–476.

Recombinant human IgG production from myeloma and Chinese hamster ovary cells

4

Ray Field

4.1 Introduction

This chapter discusses some of the current trends and issues around production of human IgG from mammalian cells for use as a therapeutic, highlighting scientific and industrial challenges to be overcome to allow efficient and cost-effective antibody therapies to be developed and marketed. A key factor to a viable process for the production of recombinant antibodies for therapy is to obtain stable high yields from cell lines at larger scales of manufacture. Factors and strategies relevant to two of the current main mammalian cell based production systems, Chinese hamster ovary (CHO) and mouse myeloma cells, will be discussed.

4.2 The need for recombinant human antibodies

The application of monoclonal antibodies to both therapy and diagnosis is now well established, with several effective marketed therapeutic antibodies generating a considerable stream of revenue (Pavlou and Belsey, 2005; Walsh, 2003). In the 1970s the first mouse monoclonal antibody was produced by cell fusion of a mouse spleen cell and myeloma cell to create a hybridoma cell (Kohler and Milstein, 1975). This pioneering work enabled a single antibody to be produced as a purified protein entity that could be characterized and produced consistently from *in vitro* cell culture in scalable quantities. However, it became apparent that the immune systems of patients treated with such mouse monoclonal antibodies produced a human anti-mouse immune response (HAMA) which limited effectiveness of the treatment. The impetus was then to create more human-like antibodies. This led to the use of human–human hybridomas (Olsson and Kaplan, 1980) and human–mouse heterohybridomas (Jessup *et al*, 2000; Leibiger *et al*, 1995) to generate human antibodies. Whilst these approaches resulted in 'human antibodies' there were several actual and potential patient safety and manufacturing implications.

(i) Extensive screening of the 'monoclonal output cells' to identify high potency monoclonal antibodies.

(ii) Tailoring of exact antibody isotype was possible but less straightforward.

(iii) Comparatively little control was possible over the stability, yield and growth characteristics of the monoclonal output cells, resulting in lower yields from manufacturing processes.

(iv) Human-derived materials were used in the cell-line construction process, leading to the increased theoretical risk of adventitious human pathogens contaminating the product. This meant that more extensive safety testing of the cell line and product was required.

4.3 Recombinant antibodies

With the advent of gene cloning techniques it became feasible to begin to address some of these shortcomings by generating human or human-like antibodies using DNA manipulation. Introducing recombinant antibody genes into cells that had capacity for increased yields and improved product safety in an industrial cell culture process also potentially improved the cost of goods. This was achieved by several different methods such as:

(i) the creation of recombinant chimeric antibodies, where a human constant domain is fused to the mouse variable domain;

(ii) the 'humanization' of mouse antibodies;

(iii) the isolation of the DNA encoding a specific human antibody gene from either a human hybridoma, spleen cell or from a genetic library of human genes.

All of these methods have their advantages and disadvantages and have been extensively reviewed (see Mountain and Adair, 1992). Ultimately, it is perceived that antibodies that are 'humanized' or human will produce less of an adverse immune response in patients and these are currently being actively developed in preference to chimeric antibodies (Brekke and Loset, 2003; Pavlou and Belsey, 2005). The chronological progression of therapeutic antibodies from mouse to human is illustrated in *Figure 4.1*.

4.4 Decoupling antibody isolation and production

The common feature of these approaches is the ability to isolate the DNA encoding the specific antibody gene of choice (recombinant antibody) so

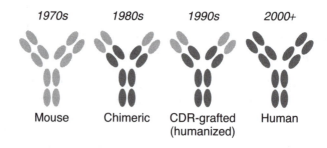

Figure 4.1

The evolution of recombinant therapeutic monoclonal antibodies from mouse to human. The lighter shades indicate more mouse amino acid sequence content, and the darker shades more human amino acid sequence.

that it can be introduced into a standardized immortal 'host' cell line (e.g. Chinese hamster ovary or mouse myeloma cells) that are suitable for production of large quantities of recombinant antibody from large culture fermentation systems. These standardized 'host cells' can then be pre-screened for desirable characteristics (e.g. growth in simple culture media, freedom from viruses, desirable genetic characteristics). This, to a large extent, decouples the cell screening required to obtain a potent antibody, from the cell screening required to obtain a high-yielding cell line. This decoupling has made a major contribution to the high titers (>5 g L^{-1} of IgG) obtained in some industrial fermenter processes of potent human antibodies that are starting to enter the marketplace today (Adamson, 2005).

One powerful approach (which we use at CAT) is to isolate human antibodies from diverse human antibody gene libraries – displayed using bacteriophage or ribosomes (Lowe, 2004; Osbourn et al., 2003). The initial antibody screening is carried out using just the binding fragments of human antibodies; for example, single-chain Fv domains (scFv), to obtain a potent scFv that binds with high affinity to a specific antigen. The DNA encoding the scFv is then easily isolated and converted to a full length (or other format) human antibody gene sequence. This is readily fused to standard, or modified, human antibody constant domain DNA sequences in a plasmid expression vector that is suitable for efficient production in Chinese hamster ovary, mouse myeloma or other cells. The use of polymerase chain reaction methods helps to standardize this whole procedure and facilitates parallel throughput and automation of the gene manipulations.

4.5 Choice of host cells

The choice of a suitable host cell to be used for introduction of the antibody gene for subsequent industrial production is limited by several factors such as:

(i)   an accepted safety profile of the cells by the regulatory agencies (e.g. FDA, EMEA);
(ii) the efficiency and ease of introduction of recombinant DNA (i.e. trans-fection);
(iii) the availability of mutants with desirable genetic characteristics that will allow the selection of efficient production of introduced genes leading to antibody protein of high yield and acceptable quality;
(iv) the ability of cells after introduction of the antibody genes to grow rapidly, in simplified defined cell culture media, and maintain a (relatively) stable genetic profile with respect to antibody production during the manufacturing process.

Currently, the most commonly used mammalian host cells are rodent cell lines such as Chinese hamster ovary cells (CHO-K1) or mouse myeloma cells (e.g. NS0, SP2/0). Rodent cell lines were chosen because they exhibited many of the desirable characteristics described above. Several human cell lines have been used as host cells (e.g. recombinant human activated protein-C is produced from human embryonic kidney (HEK) cells). Such human cell lines have the potential advantage of producing a more 'human-like' glycosylation of a recombinant antibody, although this benefit is potentially offset by the greater risk of contamination of the product with agents (e.g. viruses) that are pathogenic in human cells. More recently, a human retinoblast cell line (PER.C6) has been created that is reported to have a lower risk profile and higher yields for recombinant protein production (Jones *et al.*, 2003; Xie *et al.*, 2002).

4.5.1 Chinese hamster ovary cells

CHO cells were first isolated in the late 1950s (Puck *et al.*, 1958) and their genetic make up has been characterized extensively. Several useful mutants have been generated which allow the selection of cells expressing a selectable marker gene (linked to the recombinant antibody gene). Such mutants deficient in dihydrofolate reductase (DHFR), for example DUK-XB11 (Urlaub and Chasin, 1980) and DG44, have been widely used as a host cell for industrial production. The introduction of a plasmid containing a recombinant antibody gene and a linked *DHFR* gene into the cell followed by isolation of cells under specific nutrient culture conditions, has resulted in cell lines producing low levels of recombinant protein. Further selection of cells that have amplified copies of the introduced plasmid, by culturing cells in successively elevated levels of the DHFR inhibitor methotrexate has led to higher yielding cell lines. However, due to the toxic nature of methotrexate, the successive 'amplifications' can be quite time consuming (Kaufman, 1992). CHO cells exhibit anchorage-dependent cell growth but after adaptation are amenable to scale-up in suspension culture using simple culture media. CHO cells have ample precedent for an acceptable safety profile with the regulators (e.g. FDA) and so are often perceived as the 'host cell line of choice' for production of recombinant proteins, including antibodies (Page and Sydenham, 1991).

4.5.2 Rodent myeloma cells

Rodent myeloma cells are 'professional' secretory immune cells, with many of the desirable characteristics described above and are particularly suited to secretion of antibodies. The availability of mutants that have lost the ability to produce their own endogenous antibody (e.g. P3X63Ag8.653, SP2/0-Ag14 and NS0/1) has facilitated their use as host cells for recombinant antibody production (Yoo *et al.*, 2002). They grow naturally in suspension culture and many of the industrial manufacturing strategies previously developed using hybridoma cells (e.g. fed-batch) could be readily applied to recombinant myeloma cells, thereby speeding up process development (Biblia and Robinson, 1995). There are fewer mutant myeloma cell lines available that facilitate use of selectable marker genes (compared to CHO cells) but the more efficient secretion of antibodies from fewer gene copies compensates for this. Various strategies have been devised to ensure efficient selection of recombinant myeloma clones (Yoo *et al.*, 2002).

4.6 The glutamine synthetase system

The use of glutamine synthetase (GS) as an effective selectable marker has been described extensively elsewhere (Barnes *et al.*, 2000) and in Chapter 1 of this book. At the time of writing, the GS system is one of the few 'mammalian expression systems' that is:

(i) readily available for use in standard host cells (e.g. mouse myeloma and CHO);
(ii) has a track record creating cell lines that allow production of gram per liter titers (>5 g L^{-1}) of antibody;
(iii) does not require multiple rounds of gene amplification with associated time-consuming and labor intensive screening activities.

The glutamine synthetase gene can be used as a dominant selectable marker in many cell types. This has the added benefit that the ensuing modification of cell metabolism allows glutamine independent growth and facilitates reduced levels of toxic metabolite accumulation (e.g. lactate and ammonia; Cruza *et al.*, 2000) which leads to increased culture longevity. The GS system is most often employed by transfecting cells with a plasmid containing both the *GS* gene and antibody gene(s), followed by random integration of the plasmid into host cell chromosomes. Those recombinant cells that have 'stably' integrated the plasmid into a position in the genome that allows sufficiently high-level of expression of the recombinant *GS* gene are isolated by the appearance of cell colonies in the absence of glutamine. Such colonies often also produce significant levels of the co-integrated recombinant antibody gene. The inclusion of methionine sulfoximine (MSX), an inhibitor of GS, can facilitate this selection and is essential to suppress endogenous GS activity from host CHO cells, although not with NS0 cells. The combination of effectiveness of GS as a selectable marker and the beneficial effects of GS on cell metabolism, in many cell types including CHO and NS0 cells, has made the GS system a powerful and accessible system for industrial-scale antibody production.

4.7 Cell line stability

High yields of antibody from a manufacturing cell line require:

(i) a high specific rate of antibody production per cell (qP);
(ii) a stable high qP that is maintained after the multiple cell divisions that are needed to expand cells from initial colonies, through cell cloning, cell banking and scale up to large bioreactors;
(iii) a rapid cell growth rate, high maximum biomass and extended culture longevity under conditions of large scale industrial fermentation that also allow the high qP to be maintained.

To satisfy both regulatory and business drivers, evidence of the stability of a cell line with regard to cell growth, IgG expression level, gene copy number, transcript identity and IgG quality during large scale manufacture is required before market supply (ICH Topic Q5D).

This presents several key challenges since all of these characteristics are influenced in a variable (often random) manner by both the genetics of the cell line and by the fermentation process. Some cell lines have more inherently stable integration and expression of the introduced genes into host chromosomes, whilst other cell lines are more amenable to process manipulation than others. Appropriate screening strategies and small-scale experiments are carried out that attempt to mimic the larger scale industrial reactors to identify suitable candidates. Such experiments may examine the genetics of antibody production and the responsiveness of cell lines to feeding of nutrients and the subsequent effects on antibody secretion rate and culture longevity. Examples of such experiments are presented below.

4.8 Bioreactor process strategies

CHO and myeloma cells have been readily cultured in suspension culture and scaled up in standard stirred tank or in airlift reactors for many years (Rhodes and Birch, 1988). Developments in culture media design and understanding of cell physiology now enable these cells to be cultured in chemically defined nutrient media to achieve relatively high cell concentration and culture longevity with the consequent extended production of recombinant protein. Suspension cultures can be operated as batch or fed-batch mode (with addition of concentrated nutrient feeds), or as continuous perfusion with biomass retention (Chu and Robinson, 2001). Both strategies have their merits under different scenarios, although we have chosen fed-batch mode for production of batches of IgG with which to supply clinical trials, due to its shorter culture cycle time leading to reduced occupancy of expensive GMP fermentation capacity. The use of the GS system favours batch production, since there is some evidence that recombinant cells transfected with the GS gene, when cultured at high density in continuous perfusion, produces enough GS activity to accumulate sufficient glutamine in the culture medium to significantly reduce the effective selection pressure. This has led to the appearance of a heterogeneous cell population with lower productivity towards the end of reactor runs (Bird et al., 1998).

The driver for an 'animal component-free' (ACF) culture medium has existed for many years because of the need to reduce the potential risk of contamination of IgG product by adventitious agents such as TSEs (see Chapter 5). These agents are potentially present as contamination in fetal bovine serum (FBS), serum fractions or peptones. The goal of an efficient ACF process has now been achieved effectively for both CHO and myeloma cells. NS0 cells are a cholesterol auxotroph (Seth et al., 2005). The cholesterol requirement was previously supplied by using FBS or lipoprotein serum fractions or complexed to serum albumin, all possible TSE risks. Cholesterol can now be supplied from a non-animal source due to the advent of proprietary and commercially available preparations where non-animal sourced cholesterol is complexed to cyclodextrin (Gorfien et al., 2000; Keen and Steward, 1995). This has allowed NS0 cells to be grown in an ACF manufacturing process. In our experience, it has been easier and more efficient to carry out the entire cell line construction process for CHO cells in ACF media than for NS0 cells where some FBS supplementation was needed to maintain cell growth rates and reasonable timescales during the transfection and cell cloning stage.

4.9 IgG supply during antibody development

Typically, the quantities of recombinant antibody needed at different stages of preclinical and early clinical development vary from milligrams to hundreds of grams. A key challenge is to supply these requirements in a timely manner with IgG of appropriate quality. Smaller amounts can be readily supplied using transient gene expression from mammalian cells, where IgG expression titers of up to approximately 100 mg L^{-1} have been achieved (Prett *et al.*, 2002; Reilly, 2004) with 20–40 mg L^{-1} perhaps more common. Such transient expression using CHO or HEK-293 EBNA cells (Durocher *et al.*, 2002; Prett *et al.*, 2002) allows efficient production of milligram to hundred milligram quantities of multiple IgGs in relatively short timescales (weeks). Where transient expression has been efficiently scaled to 10-L or 100-L reactors, then multi-gram amounts can be rapidly produced (Reilly, 2004). One important factor is the comparability of this 'early stage' IgG to the final product, particularly with regard to the impact of variation in glycosylation of the IgG on its potency. Whilst HEK-293 cells are efficient for transient gene expression, they are a human-derived cell line and so impart a different glycoform profile to CHO or NS0 cells (Jefferis, 2001). This may be significant to the antibody development program, depending on how much the effector function of the antibody contributes to its mode of action *in vivo*, since variation in antibody glycosylation has most impact on effector function (Jefferis, 2001). If the mode of action of the recombinant IgG is based predominantly on binding to an antigen, then variation in IgG glycoform between early supplied IgG and clinical development candidate may be of less importance.

To start to understand the potential efficacy of a therapeutic IgG *in vivo*, gram amounts of several candidate molecules are often needed for pharmacology studies. It is at this point where a decision between scale-up of transient expression and production of stable or semi-stable cell lines often needs to be taken. It can ultimately be less resource intensive to use a stable cell line since this facilitates repeated production batches from the cell line stock, rather than having to carry out multiple repeated transient transfections at large scale for each batch. Often the timing of the requirement for such quantities of IgG during development allows early demands to be supplied from transient expression, meanwhile transfections for stable cell lines are initiated. So by the time pharmacology studies require larger multi-gram quantities of IgG, a sufficiently productive nonclonal stable cell line is usually available.

4.10 Strategies for cell line engineering during clinical development

To supply IgG for human clinical trials, ideally a clonal, stable, high-yielding cell line is preferred. Generating a suitable cell line can be a resource and time consuming activity. A judgement of whether to adopt the opposing strategies of 'obtain the best cell line you can' from the start of early clinical development, or obtain a cell line 'just good enough then change it later' needs to be made. The former strategy takes longer and has higher costs to supply IgG to start clinical development with the risk of an unjustified investment should the product then fail in the clinic. However, it has the benefit that the initial GMP production batches would be at a smaller volumetric scale and supply of IgG for clinical development to phase III (pivotal studies) and beyond would be accelerated with less likelihood that any noncomparability of product between early and pivotal clinical studies would cause a hiatus in the development program. The latter strategy would reduce the time and cost of supply of early clinical trials material with the risk that a new cell line may be needed for later clinical development. This may incur significant time delays, extra costs and uncertainty of product comparability as the program heads towards phase III clinical trials.

Fortunately, the GS system allows fairly rapid development of high-yielding cell lines. Our current strategy at CAT falls between the two scenarios. We can achieve a 1–3 g L^{-1} titer from an 'early clonal cell line' and believe that this can be improved by subsequent process development to achieve even higher yields without re-engineering the cell line or laying down a new master cell bank (MCB). This is the case both with GS–NS0 and GS–CHO cells.

4.11 Cost of goods and intellectual property

A significant factor in the route to manufacture of IgG is adequate consideration of the predicted cost of goods at anticipated final production scale. This is a complex series of calculations that must take into account the actual costs of manufacture (e.g. raw materials, labor costs, capital costs), the process yield, scale of operation and also any license fees that may be due as a result of third-party patents. Within the antibody development and manufacturing arenas this has become a complex issue with patents that could impact on the discovery technology, the vector expression system and host cells used for manufacture, and various other aspects of the production process. Obviously, the details of any financial implications of the need to license such third-party patents are commercially sensitive and are typically kept confidential. However, these issues are becoming more of a driver towards different routes of antibody manufacture. For example, the current intellectual property landscape with regard to the use of either myeloma cells or CHO cells for recombinant antibody manufacture is quite different. Furthermore, the current patent situation in the more general antibody manufacturing arena differs between Europe and the US and is fluid as a result of various patent oppositions and litigation, adding a reasonable degree of uncertainty as to the final 'cost of goods', depending on exactly when the IgG product actually comes to market.

4.12 Recombinant human IgG production from myeloma and CHO cells

It has been pertinent to compare mouse myeloma NS0 with CHO cells for the production of recombinant human IgG to determine any practical advantages of each system from a perspective of productivity, timelines scalability and resources used for cell line construction and process development. The GS system has been used to obtain high levels of IgG expression and comparisons of cell line stability and performance under fed-batch bioreactor conditions are most relevant.

4.12.1 Creation of CHO and NS0 cell lines expressing IgG

The GS system licensed from Lonza Biologics is used to create GS–CHO and GS–NS0 cell lines that express recombinant human IgG. The transfection of NS0 or CHO-K1 suspension variant (CHOK1SV) cells uses a plasmid expression vector into which have been cloned a human IgG heavy and light chain gene in tandem. In summary, cells transfected by electroporation are selected in glutamine-free medium in 96-well plates. CHO cells are selected in the presence of 50 µM MSX. After approximately 3–4 weeks, supernatants are sampled from wells containing surviving colonies are assayed for levels of IgG using a sandwich ELISA that detects assembled heavy and light chain. The highest producing colonies, expanded to 24-well plates are assayed for levels of secreted IgG by HPLC-protein-A (Dempsey *et al.*, 2003). This process is continued until candidate cell lines are adapted to suspension culture in serum-free medium in Erlenmeyer flask culture. The best cell lines at this stage can then be chosen for dilution cloning in 96-well plates and the subsequent clones expanded and screened as before. Ultimately, clonal cell line candidates are adapted to serum-free medium and research cell stocks (designated as generation number zero) are cryopreserved in liquid nitrogen. Ampoules of the research cell stock are revived and cells characterized for cell line stability over long-term subculture. A separate ampoule from the research stock is subsequently used to generate a GMP master and working cell bank (MCB and WCB) to supply IgG for clinical trials.

4.12.2 Cell expansion, subculture and production reactor experiments

Viable cell concentrations are analyzed by trypan blue exclusion using a CEDEX automated cell counter. NS0 cells are cultured in suspension using in-house proprietary 'GSF' serum-free media and CHO cells are cultured in suspension using CD-CHO medium (Gibco). Glutamine-free conditions are maintained at all times and CHO cells also supplemented with 25 µM MSX during routine subculture. Cells are subcultured regularly at fixed time-intervals and then seeded at 2×10^5 cells mL^{-1} by dilution with fresh culture medium. Scale-down flask experiments and routine subcultures are carried out in sterile disposable Erlenmeyer flasks incubated at 36.5°C on an orbital shaker base. In-house proprietary nutrient feeds are added to cultures using a regime based on analysis of nutrient consumption rates in a fed-batch mode (Dempsey *et al.*, 2003). Bioreactor production is carried out using suspension cultures in Applikon stirred tank reactors where temperature, dissolved oxygen, pH and stirrer speed are controlled (Dempsey *et al.*, 2003).

4.12.3 Northern and western blotting

Northern blots carried out using digoxigenin (DIG) labeled probes using a Roche PCR DIG probe synthesis kit are designed to anneal to the heavy or light chain variable regions of the IgG sequence. Analysis is carried out using samples of total RNA extracted from 1×10^7 cells. Western blots can be probed using polyclonal antisera raised against human heavy chain Fc domain or human light chain constant domain.

4.12.4 Comparison of results of transfections from GS–NS0 and GS–CHO

Transfection of both NS0 and CHOK1SV cells results in colonies producing a wide distribution of IgG concentration. Large numbers of transfectants

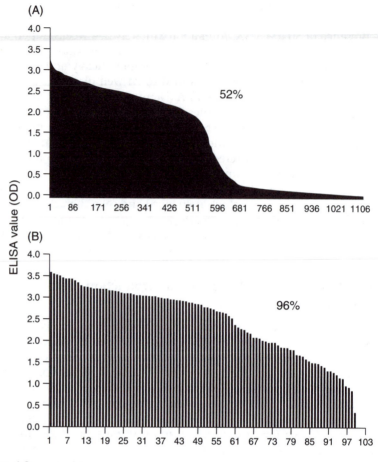

Figure 4.2

The relative distribution of IgG-producing colonies from GS-NS0 (A) and GS-CHO (B) transfections. The Y-axis shows the concentration of IgG produced, displayed as an absorbance value (OD) from a sandwich ELISA. The X-axis shows in rank order the number of colonies assayed from 96-well plates. The percentage figure is the proportion of colonies with an ELISA OD value >1.0

can be produced from NS0 transfections, whilst substantially fewer transfectants arise from CHO transfections. ELISA assay of supernatants from these colonies indicates there is a higher proportion of CHO cells producing elevated levels of IgG when compared to NS0 cells, as shown in *Figure 4.2*.

4.12.5 Dilution cloning and analysis of clonal heterogeneity

After analysis of productivity, the highest producing transfectants are selected for dilution cloning in 96-well plates. Probability of monoclonality can be calculated by an extended form of a method originally based on Poisson distribution analysis (Coller and Coller, 1986). As part of this assessment procedure for GS–NS0 cells the best candidate clones obtained from

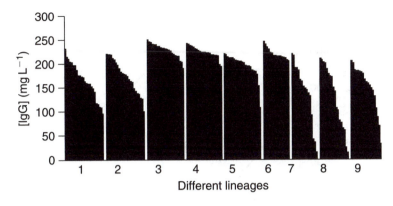

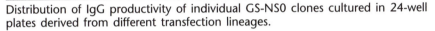

Figure 4.3

Distribution of IgG productivity of individual GS-NS0 clones cultured in 24-well plates derived from different transfection lineages.

each transfection lineage are cultured in 24-well plates and the level of IgG accumulation analyzed. *Figure 4.3* shows the typical heterogeneity of IgG production of clones arising from several transfection lineages. Those transfection lineages that gave rise to clones showing the most homogeneous IgG production are deemed most suitable for further progression.

4.12.6 Analysis of instability of a GS–NS0 cell line

Despite various screening procedures, unstable GS–NS0 cell clones do still arise at a frequency of less than 50%. The problem of instability manifests itself by a progressive decrease in IgG production after an extended period of subculture. Cell line instability can be a potential issue since expansion of the seed stock from master cell bank vial through to the final production scale reactors can require 40–50 cell doublings for large scale manufacture. Racher (2005), has defined instability as a greater than 30% decrease in cell specific production rate between the cell bank and the defined limit of *in vitro* cell age. This is a useful definition, although in practice the significance of the cut off (e.g. 30% decrease) will vary depending on circumstances such as the absolute expression titre. For example, in our experience at CAT, cell line stability can vary markedly between cell lineages expressing different antibodies. An example of a stable and severely unstable NS0 cell clone is shown in *Figure 4.4*.

In order to better understand at the genetic level the mechanism of instability in NS0 cells, culture supernatants from GS–NS0 clones can be analysed by western blotting using probes specific to immunoglobulin sequences and also by northern blotting to determine relative levels of heavy and light chain transcripts. Results shown in *Figure 4.5*, suggest that instability of IgG production coincides with reduced levels of immunoglobulin heavy chain protein. Reduced levels of immunoglobulin heavy chain transcripts determined by northern blots generally do correlate with this (*Figure 4.6*). In contrast, the example shown in *Figures 4.5 and 4.6*, shows

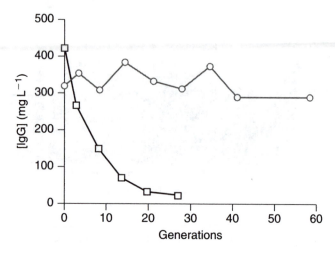

Figure 4.4

Productivity of a stable and unstable NS0 cell clone after extended serial subculture in flasks. Harvest titer from saturated supernatants is plotted on the Y-axis and the calculated cell generations are plotted on the X-axis. (O), Stable cell line; (□), unstable cell line.

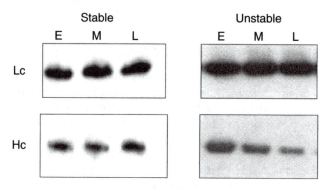

Figure 4.5

Western blot of heavy (Hc) and light (Lc) chains from culture supernatant taken from early (E), middle (M) and late (L) generation stable and unstable NS0 clones. Instability is associated with decreased heavy chain secretion at higher generation numbers.

that light chain protein and transcripts remain constant. This is typical of the pattern of instability seen with recombinant NS0 cells producing IgG.

4.12.7 Output of transfections of GS–NS0 and GS–CHO

A comparison of the output of the transfections of GS–NS0 and GS–CHO with the same recombinant human antibody is shown in *Table 4.1*, summarizing the timelines and characteristics of the lead cell lines obtained. The

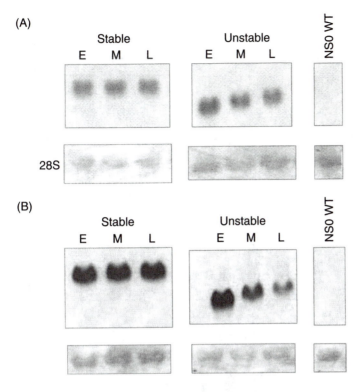

Figure 4.6

Northern blot of heavy or light chain human immunoglobulin transcripts from stable or unstable NS0 clones. 28S ribosomal RNA is indicated as a loading control. NS0 WT is a negative control from untransfected NS0 cells. (A) Light chain; (B) heavy chain.

Table 4.1 Summary of output from GS-NS0 and GS-CHO transfections

	GS-NS0	GS-CHO
Percent of colonies producing IgG	50–60	96
Time (weeks) to serum-free culture	18	16
Number of colonies screened	1120	102
Productivity (in mg L^{-1}) of lead stable cell line (unfed)	999	796

number of colonies obtained for GS–NS0 is approximately 10-fold greater than GS–CHO. However, the productivity of the lead stable cell lines are similar, at this stage, for both CHO and NS0.

4.12.8 IgG production stability of candidate GS–NS0 clones

GS–NS0 production clones can be evaluated for stability of IgG production over an extended period, cultured in glutamine-free animal component-free medium without MSX addition to simulate expansion to large scale production volumes. *Figure 4.7A* schematically represents the stability assessment

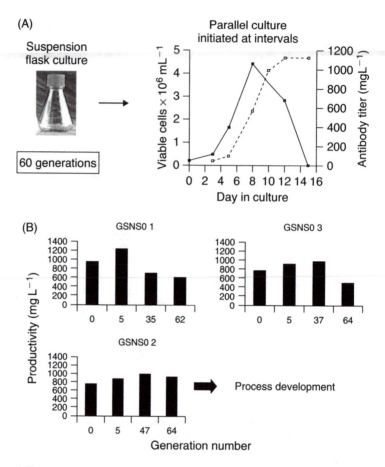

Figure 4.7

(A) Stability assessment of cell lines in flask culture. Suspension cultures were subcultured using a regime designed to simulate inoculum expansion to a bioreactor. At intervals parallel cultures were allowed grow to saturation and the IgG harvest titer was assayed by HPLC-protein-A. (B) Stability study of the lead GS-NS0 clones over extended subculture. IgG productivity was measured by HPLC-protein-A assay.

in flask cultures and *Figure 4.7B* shows typical results obtained from three representative GS–NS0 clones.

4.12.9 IgG production stability of GS–CHO transfectants

Figure 4.8 shows results from analysis of production stability of the six highest-producing GS–CHO transfected parental cell lines (i.e. prior to dilution cloning) subcultured for an extended number of population doublings in the presence of 25 µM MSX. Although these parental cell lines are not 'clones', the low number of colonies arising from the transfection effectively ensures that the cell lines are essentially clonal. At intervals, parallel flasks are grown to saturation to measure the volumetric concentra-

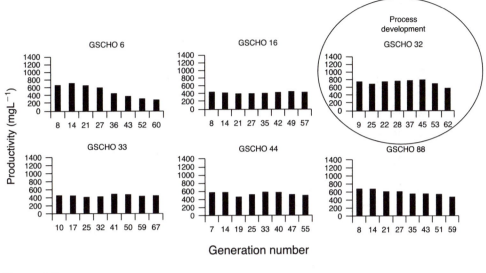

Figure 4.8

Stability of IgG production from GS-CHO cell lines during extended flask culture. The cell line GSCHO32 was chosen for further process development.

tion of IgG in supernatant. Five of the six cell lines show high levels of stability of IgG production.

4.12.10 Fed-batch bioreactor process for GS–NS0 and GS–CHO

The most suitable GS–NS0 and GS–CHO cell lines are chosen on the basis of performance in flask cultures, response to feed, growth rate and cell line stability. Scale-up to bench-top stirred tank reactors is then carried out using animal component free media. NS0 cells can be cultured in a CAT proprietary medium GSF-14 and CHO cells cultured in a commercially available CD-CHO medium. Bioreactors are controlled as described above without MSX selection in the production reactors.

Figure 4.9 shows results of initial fed-batch fermentations of GS–CHO and GS–NS0 cells. The NS0 process has undergone extensive iterations of nutrient analysis and subsequent modification of the fed batch regime, compared to the CHO process. However, since both CHO and NS0 processes yielded in excess of 2 g L^{-1} IgG, this indicates that such nutrient feeding strategies account for the modified cell metabolism after transfection of cells with the *GS* gene, have some similarity between CHO and NS0.

4.12.11 Analysis of IgG quality produced from GS–CHO and GS–NS0 bioreactor processes

The recombinant antibody produced from cultures and purified by protein-A chromatography is suitable for analysis by mass spectrometry, SDS-PAGE, isoelectric focussing profile and an *in vitro* potency assay. Such results show

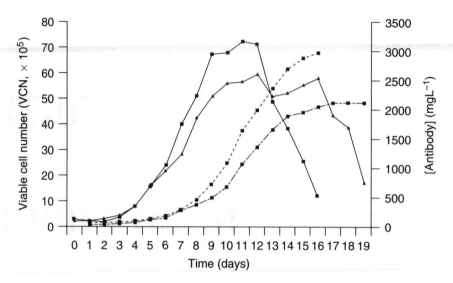

Figure 4.9

Comparison of growth and IgG accumulation of fed-batch reactor cultures of lead GS-CHO and GS-NS0 cell lines producing the same human IgG.
—■— NSO VCN; —▲— CHO VCN; – –■– – NSO (Ab); — —■— — CHO (Ab).

(A)

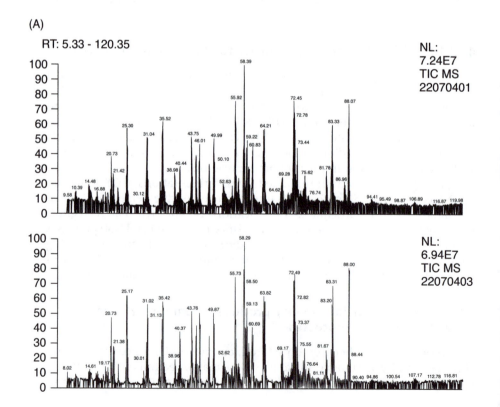

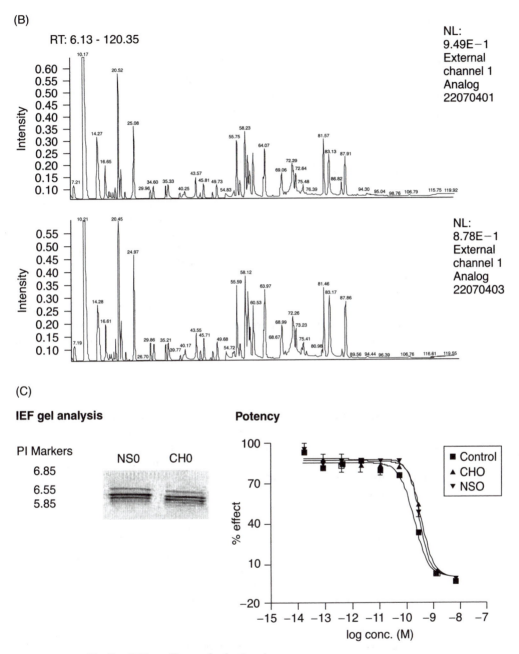

Figure 4.10

(A) Electrospray mass spectrometric analysis of reduced tryptic digest of purified human IgG. Total-ion maps of human IgG from NS0 (top) and CHO (bottom) cells. (B) Mass spectrometry UV chromatograms of the isolated oligopeptide containing oligosaccharide from NS0 (top) or CHO cells. (C) Comparison of human IgG produced from NS0 and CHO cells by isoelectric focussing gel analysis and an *in vitro* cell-based potency assay.

that the IgG produced from both CHO or NS0 are generally comparable (as shown in *Figure 4.10*), although some minor changes in physical characteristics would be expected due to the differences in terminal glycosylation of the antibody between CHO and NS0. Any difference in glycoform of an antibody generally has most impact upon the effector function mediated by the Fc domain (Jefferis, 2001). In this particular example the mode of action of the human IgG4 antibody is not thought to require effector function so minor glycoform differences would not impact upon an *in vitro* potency assay.

4.12.12 Comparative yield of different human IgGs produced from CHO or NS0 cells

Process development for antibody production from GS–CHO and GS–NS0 cells is at a stage where the bench-top bioreactor production titers from either system in a bioreactor process that is free of animal components are similar. The GS–NS0 system has currently a more extensive track record of scale-up and more antibodies have currently been produced from the GS–NS0 system compared to GS–CHO. However, the early indication is that the GS–CHO system (using CHOK1SV) is amenable to bioreactor process development and rapid progress has been possible to achieve similar volumetric productivity compared to the GS–NS0 system. A summary of representative bioreactor production titers from GS–NS0 and GS–CHO for several human IgGs are shown in *Figure 4.11A and B*.

4.13 Summary

The production of recombinant human monoclonal antibodies from mouse myeloma and CHO cells is now well precedented. However, due to the amount of extensive process development required, it is only occasionally that the two expression and production systems are compared using commercially relevant immunoglobulins and procedures. Procedures for transfection of CHO cells resulted in high-level IgG production from most of the colonies resulting from transfection and generated cell lines with a high level of long term stability during extended subculture. Typically, at least 50% of clonal NS0 cell lines are stable. Analysis of instability of IgG production showed that IgG productivity decrease can correlate with decreased production of heavy-chain protein and transcript levels, with light chain production remaining constant. This information allows a cell line screening and development strategy for GS–NS0 cells focussing on heavy chain transcript levels to be implemented. Instability of NS0 recombinant IgG expression has been investigated by others (Barnes *et al.*, 2004; Racher, 2005) in addition to our own work (Daramola *et al.*, 2005). In the work reported by Barnes *et al.* (2004), absolute levels of heavy- and light-chain IgG transcripts were predictive of phenotypic expression stability provided they were above a threshold level. Our work, and that of Racher (2005), also supports a metabolic effect on instability which is potentially reversible. Hence, the exact nature of culture conditions may lead to differences in observations of frequency and magnitude of cell line instability in GS–NS0 cells. Nevertheless, it is clear that stable GS–NS0 cell lines can be selected with relative ease and abundance.

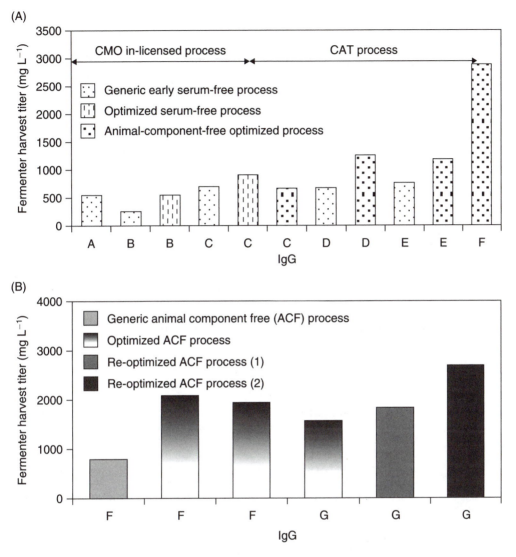

Figure 4.11

(A) Harvest titer improvements in bioreactor processes for GS-NS0 cells producing different human IgGs, (B) Harvest titer improvements in bioreactor processes for GS-CHO cells producing different human IgGs.

Comparison of GS–CHO and GS–NS0 cell lines and bioreactor processes for production of multiple human IgGs shows that both systems are capable of producing multi-gram per liter volumetric titers of human IgG. Extensive process development of GS–NS0 cells was required to achieve a high-yielding animal component-free bioreactor process, whereas this was established more rapidly for GS–CHO cells. It is envisaged that since the use of the GS–CHO system has already rapidly lead to high IgG expression titers, then this system will be widely adopted for antibody production on the basis that further benefits on productivity should also be achieved in the near future.

Acknowledgments

The author wishes to acknowledge the invaluable contributions of the Cell Sciences group at CAT (Alison Ridley, Chris Tape, Diane Hatton, Gareth Lewis, Greg Dean, Janice Myers, Jessica Prett, Jonathan Dempsey, Kalpana Nayyar, Karen Maisey, Lekan Daramola, Steve Ruddock, Steven Doig, Suzanne Gibson, Tori Crook, Wyn Forrest-Owen), members of the Cell Process Development, Cell Engineering, Antibody Format Technologies, Bioprocess Sciences, Analytical and Quality Control teams at CAT who generated the data presented here. Thanks to Sandrine Mulot for providing the data for, and help with Figure 4.10. Also to Lonza Biologics for advice on the GS system. Thanks to Jon Dempsey, Diane Hatton and Lekan Daramola for review and comments of the manuscript.

References

Adamson R (2005) Technologies employed to manufacture currently available biotechnology products: Current considerations and future challenges. Presentation at IBC's 2nd Early Development of Biotherapeutics Congress, 25–27 April 2005, Reston VA. Biopharmaceutical Production Series, available from Informa Life Sciences group, London or http://ibclifesciences.com/handbooks.

Barnes LM, Bentley CM and Dickson AJ (2000) Advances in animal cell recombinant protein production in the GS-NS0 expression system. *Biotechnol Bioeng* **73**: 261–270.

Barnes LM, Bentley CM and Dickson AJ (2004) Molecular definition of predictive indicators of stable protein expression in recombinant NS0 myeloma cells. *Biotechnol Bioeng* **85**: 115–121.

Biblia TA and Robinson DK (1995) In pursuit of the optimal fed-batch process for monoclonal antibody production. *Biotechnol Prog* **11**: 1–13.

Bird P, Bolam E, Castell L, Obeid O, Darton N and Hale G (1998). Glutamine synthetase transfected cells may avoid selection by releasing glutamine. In: *New Developments and New Applications in Animal Cell Technology*. (eds O.W. Merten, P. Berrin and B. Griffiths). Kluwer Academic Publishers, pp. 43–49.

Brekke OH and Loset GA (2003) New technologies in therapeutic antibody development. *Curr Opin Pharmacol* **3**: 544–550.

Chu L and Robinson DK (2001) Industrial choices for protein production by large-scale cell culture *Curr Opin Biotechnol* **12**: 180–187.

Coller HA and Coller BS (1986) Poisson statistical analysis of repetitive subcloning by the limiting dilution technique as a way of assessing hybridoma monoclonality. *Meth Enzymol* **121**: 412–417.

Cruza HJ, Freitasa CM, Alvesa PM, Moreiraa JL and Carrondo MJT (2000) Effects of ammonia and lactate on growth, metabolism, and productivity of BHK cells. *Enzyme Microb Technol* **27**: 43–52.

Daramola O, Hatton D and Field R (2005) Meeting the challenges of IgG expression: from antibody libraries to clinical supply. *Bioprocess J* **4**: 33–37.

Dempsey J, Ruddock S, Osborne M, Ridley A, Sturt S and Field R (2003) Improved fermentation processes for NS0 cell lines expressing human antibodies and glutamine synthetase. *Biotechnol prog* **19**: 175–178.

Durocher Y, Perret S and Kamen A (2002) High-level and high-throughput recombinant protein production by transient transfection of suspension-growing human 293-EBNA1 cells. *Nucleic Acids Res* **30**: e9.

Gorfien S, Paul B, Walowitz J, Keem R, Biddle W and Jayme D (2000). Growth of NS0 cells in protein-free, chemically defined medium. *Biotechnol Prog* **16**: 682–687.

International Conference on Harmonisation. ICH Topics Q5D – Derivation and Characterisation of Cell Substrates Used for production of Biotechnological/ Biological products. Step 4 consensus guidelines 16 September 1997 (CPMP/ ICH/294/95). ICH Q5D is published in the Federal Register, Vol 63, No 182, September 21, 1998, pages 50244–49.

Jefferis R (2001) Glycosylation of human IgG antibodies: relevance to therapeutic applications. *BioPharm* September: 19–27.

Jessup CF, Baxendale H, Goldblatt D and Zola H (2000) Preparation of human-mouse heterohybridomas against an immunising antigen. *J Immunol Meth* **246**: 187–202.

Jones D, Kroos N, Anema R *et al.* (2003) High-level expression of recombinant IgG in the human cell line PER.C6. *Biotechnol Prog* **19**: 163–168.

Kaufman RJ (1992) Amplification and expression of transfected genes in mammalian cells. In: *Gene Amplification in Mammalian Cells,* (ed R.E. Kellems). Marcel Dekker Inc. pp. 315–344.

Keen MJ and Steward TW (1995). Adaptation of cholesterol-requiring NS0 mouse myeloma cells to high density growth in a fully defined protein-free and choles- terol-free culture medium. *Cytotechnology* **17**: 153–163.

Kohler G and Milstein C (1975) Continuous cultures of fused cells secreting antibody of predefined specificity. *Nature* **256**: 495–497.

Leibiger H, Hansen A, Schoenherr G, Seifert M, Wuster D, Stiger R and Marx U (1995) Glycosylation analysis of a polyreactive human monoclonal IgG antibody derived from a human-mouse heterohybridoma. *Mol Immunol* **32**: 595–602.

Lowe D (2004) Antibody engineering for future therapeutics. *Biotechnol Int* **16**: 8–13.

Mountain A and Adair JR (1992) Engineering antibodies for therapy. *Biotechnol Genet Eng Rev* **10**: 1–142.

Olsson L and Kaplan HS (1980) Human-human hybridomas producing monoclonal antibodies of predefined specificity. *Proc Natl Acad Sci USA* **77**: 5429–5431.

Osbourn J, Jerumutus L and Duncan A (2003) Current methods for the generation of human antibodies for the treatment of autoimmune diseases. *Drug Discovery Today* **8**: 845–851.

Page MJ and Sydenham MA (1991) High level expression of the humanized monoclonal antibody Campath-1H in Chinese hamster ovary cells. *Biotechnology* **9**: 64–68.

Pavlou AK and Belsey MJ (2005) The therapeutic antibodies market to 2008. *Eur J Pharm BioPharm* **59**: 389–396.

Prett J, Daramola O, Cohen M, Davies S, Field R and Hatton D (2002) Rapid produc- tion of IgG from scFv. Poster presented at the Second European Biotechnology Workshop, 15–17 September 2002, Kartause Ittingen, Switzerland. Poster avail- able on request from info@cambridgeantibody.com

Puck TT, Cieciura SJ and Robinson A (1958) Genetics of somatic mammalian cells III. Long-term cultivation of euploid cells from human and animal subjects. *J Exp Med* **108**: 945–956.

Racher A (2005) Stability and suitability of GS-NS0 cell lines for manufacturing antibodies. *Bioprocess J* **4**: 61–65.

Reilly D (2004) Rapid production of proteins with large-scale transient transfection cultures. Presentation at IBC's Cell Culture and Upstream Processing Conference, 13–14 September 2004, Berlin. Biopharmaceutical Production Series, KF1015, available from Informa Life Sciences group, London.

Rhodes M and Birch J (1988) Large-scale production of proteins from mammalian cells. *Biotechnol* **6**: 520–523.

Seth G, Philp RJ, Denoya CD, McGrath K, Stutzman-Engwall KJ, Yap M and Hu WS (2005) Large-scale gene expression analysis of cholesterol dependence in NS0 cells. *Biotechnol Bioeng* **90**: 552–567.

Urlaub G and Chasin LA (1980) Isolation of Chinese hamster cell mutants deficient in dihydrofolate reductase activity. *Proc Natl Acad Sci USA* **77**: 4216–4220.

Walsh G (2003) Biopharmaceutical benchmarks – 2003. *Nat Biotechnol* **21**: 865–870.

Xie L, Pilbrough W, Metallo C, Zhong T, Pikus L, Leung J, Aunins J and Zhou W (2002). Serum-free suspension cultivation of PER.C6 cells and recombinant adenovirus production under different pH conditions. *Biotechnol Bioeng* **80**: 569–579.

Yoo EM, Koteswara RC, Penichet ML and Morrison SL (2002) Myeloma expression systems. *J Immunol Meth* **261**: 1–20.

Media development

Cell culture media development: Customization of animal origin-free components and supplements

5

Stephen F. Gorfien

5.1 Introduction

Variability in nutritional requirements of commonly used production cells points to a need for culture media and supplements tailored to the specific cell line and process. The customization process can be complicated by concerns regarding the safety of the raw materials. Regulatory agency guidance on raw materials used in the production of human biotherapeutics has prompted an evolving definition of what constitutes an animal-derived component. Traceability of raw materials further back in the supply chain is now a requirement and non-animal origin components may be unacceptable if a secondary manufacturing process involving animal-derived material was employed. This chapter traces recent developments in sourcing of non-animal replacements. Benefits as well as potential problems associated with non-animal derived components are discussed.

Traditional cell culture techniques were developed to establish and grow a variety of cells *in vitro* for purposes of studying genetics and basic cellular processes. The original methods, culture media and supplements designed for cell growth are more than 50 years old and never anticipated modern applications for continuous cell lines. The evolution of cell culture media and processes has been an iterative process, each new development building on the foundations of earlier work. The pioneering work of Eagle (1955), Ham (1965), Moore *et al.* (1967), Barnes and Sato (1980) and others (for review, see Jayme *et al.*, 1997) formed the basis upon which many of today's culture media for biological production were designed. The range of different cell types used for biological production of recombinant proteins, viruses and viral vectors is very broad (see *Table 5.1*). The species of origin, anatomical site of origin, culture/clonal history and in the case of transfected cells, even the site of integration of foreign DNA can have an impact on the physiology of a cell line and therefore may affect the nutritional requirements of the culture. The type and scale of the culture process may

Table 5.1 Cell lines used for biological production

Cell line	Examples	Application		
		Rec protein	Virus	Gene therapy
Chinese hamster ovary (CHO)	CHO DG44, CHO DUKXB-11, CHO-K1	X		
Murine myelomas	NS1, NS0, Sp2/0	X		
Hamster kidney	BHK21	X	X	
Human kidney	293F, 293H, 293T,	X	X	X
Human retinoblast	PER.C6®	X	X	X
Canine kidney	MDCK		X	
Bovine kidney	MDBK		X	
Porcine kidney	PK15		X	
Primate kidney	Vero		X	
Human lung	MRC-5		X	
Human lung	WI-38		X	
Human cervical carcinoma	HeLa		X	
Invertebrate (insect)	Sf9, Sf21, Tn5B1-4	X	X	
Primary cells	CEF		X	

cause variation in nutrient utilization, and variability in the raw materials used to produce the culture medium may also affect biological performance attributes of the culture. These factors must be taken into account during the development and optimization of a culture medium and overall process. This chapter describes the types of culture media available for biological production and issues surrounding media components.

5.2 Types of cell culture media

Classical cell culture media are comprised minimally of water, salts, amino acids, vitamins, nucleic acid precursors, and carbohydrates. Depending on the application, supplements are often added, the most common one being animal-derived serum. Fetal bovine serum in particular, is commonly used as a supplement to classical media (Gruber and Jayme, 1994; Jayme and Gruber, 1994). Sera and other animal-derived components are still used in many biological production processes, but issues of definition, safety, consistency of performance, and cost have contributed to the development of serum-free alternatives (Jayme, 1999; Merten, 1999). Serum-free media (SFM) are therefore a subset of all media, in which cells are capable of expanding and/or expressing a product in the absence of serum supplementation (*Figure 5.1*). To eliminate the use of serum, it is necessary to understand the many functions of serum, which include binding/carrier functions for lipids, hormones, growth factors, attachment factors, cytokines, and trace metals, as well as bulk protein functions like protease inhibition, shear protection, and buffering (Jayme and Blackman, 1985). Early serum-free media often contained one or more serum fractions like albumin or fetuin, in addition to insulin, transferrin, a source of lipids (often an animal-derived lipoprotein preparation), trace metal salts, and elevated levels of amino acids, vitamins, and carbohydrates compared to classical media. Cells often required a period of adaptation to SFM and it soon became apparent that a 'universal SFM' did not exist. As SFM came to

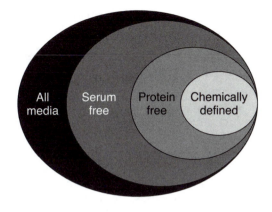

Figure 5.1

Different types of culture media. Chemically defined media are a subset of protein-free media which are a subset of serum-free media which are a subset of all culture media. The GIBCO definition of chemically defined also means that the formulation is free of any components of direct animal origin.

be used more frequently for biological production applications, it also became apparent that the undefined nature of some of the medium constituents represented a potential cause of culture performance variability. Of greater concern was the potential for transmission of adventitious agents from the animal-derived medium components to patients receiving a biological therapeutic or prophylactic that had been produced in cultured cells. These factors have led to the development of protein-free media (PFM) and chemically defined media (CDM). PFM is a subset of SFM (*Figure 5.1*). It should be noted that there are many commercially available sources of cell culture media and not all suppliers use the same definitions. For purposes of the examples provided in this chapter, PFM means that there are no polypeptides greater than 10 kDa in the formulation, although it is possible for the formulation to contain very low molecular weight peptides as might be contributed by protein hydrolysates (mainly di- and tri-peptides) that have been subjected to ultrafiltration. A CDM is a formulation in which the chemical structure of each component is known. This is another potential source of confusion among different manufacturers of cell culture media. Some manufacturers consider media that contain proteins like insulin or transferrin to be chemically defined while others' definitions require that a CDM must also be protein-free. All examples provided in this chapter consider CDM to be a subset of PFM (*Figure 5.1*).

5.3 Components of animal origin

As media manufacturers became more attuned to the regulatory requirements of their customers working in biological production, they began to investigate the source of the raw materials used in the manufacture of cell culture media and reagents. It quickly became apparent that there were some components in many media formulations that were of direct animal origin. Insulin and transferrin were already recognized as being of animal origin (although recombinant insulin has been available for some time), but it was learned that some amino acids were obtained from human hair, avian feathers or bovine gelatin. A major effort was undertaken to source animal origin-free media and reagent components and this effort uncovered an interesting fact. Tracing back the supply chain for raw materials showed that for many components not of direct animal origin, a process may have been used in the manufacture of that raw material that contained an animal-derived product. An example of this type of situation is a raw material made by microbial fermentation employing a meat digest in the fermentation broth. This raises the question of what is the definition of animal origin. Material *directly* derived from animal tissues, cells or body fluids of eukaryotic organisms such as mammals (including humans), fish, birds, reptiles, amphibians, insects, molluscs, etc is the definition of animal origin that will be used in this chapter, although it should be recognized that there is some debate as to whether material from nonmammalian eukaryotes represent the same level of risk as material derived from mammals (and therefore whether nonmammals should be considered acceptable for biological production applications). The definition of animal origin *does not* include eukaryotic organisms such as the higher plants, fungi, protozoa and algae, nor does it include prokaryotic organisms such as bacteria or blue–green algae. The definition is evolving and in some geographies, it includes materials made from a process in which animal derived components have been used.

The appearance of bovine spongiform encephalopathy (BSE) was a major factor in the move away from animal origin raw materials in the manufacture of human biotherapeutics. EMEA/410/01 rev 2 (October, 2003) describes the European Scientific Steering Committee classification for geographic BSE risk (GBR) (*Table 5.2*).

The same guidance document also groups tissues into three major infectivity categories, irrespective of the stage of disease (*Table 5.3*). Categories in the tables are only indicative and it is important to note that cross-contamination potential may exist through contact of low-risk material with high-risk material. EMEA has stated that where available, use of materials from 'nonTSE relevant species' or non-animal origin is preferred (EMEA/410/01 rev 2). According to EMEA, minimizing risk of TSE should be based upon three complementary parameters:

(i)   source animals and geographical origin;
(ii) nature of animal material and any procedures to avoid cross contamination with higher risk materials;

Table 5.2 GBR classification for BSE risk

GBR level	Presence of one or more cattle clinically or preclinically infected with BSE in a geographical region/country
I	Highly unlikely
II	Unlikely, but not excluded
III	Likely, but not confirmed, or confirmed at a lower level
IV	Confirmed at a higher level*

*>100 cases per 1 million adult cattle per year.

Table 5.3 Tissue infectivity categories

Category	Description
A	High infectivity tissues: Central nervous system (CNS) tissues that attain a high titer of infectivity in the later stages of all TSEs, and certain tissues that are anatomically associated with the CNS
B	Lower infectivity tissues: peripheral tissues that have tested positive for infectivity and/or PrPSc in at least one form of TSE
C	Tissues with no detectable infectivity: tissues that have been examined for infectivity, without any infectivity detected, and/or PrPSc with negative results

The tissue classifications are based on the 'WHO guidelines on transmissible spongiform encephalopathies in relation to biological and pharmaceutical products' (February 2003) WHO/BCT/QSD/03.01.

(iii) production process(es) and QA system to ensure product consistency and traceability.

A sensible approach to this issue from a medium manufacturing perspective includes the following strategies:

(i)   segregate,
(ii) mitigate,
(iii) replace.

5.3.1 Segregate

Raw materials are often sourced from suppliers with facilities located in diverse geographic locations. Some areas may be endemic for a particular adventitious agent and therefore represent a greater risk than an isolated geographic area in which the disease-causing organism has not been demonstrated. Certain animal species may be more susceptible to a specific adventitious agent. Increased risk may be associated with specific anatomic sites within a single animal. Sourcing of animal derived raw materials from geographic areas of low risk is preferred if an animal origin component must be used. Similarly, use of raw materials from lower risk species, organ systems, tissues or body fluids can help to reduce concerns over transmission of adventitious agents when there are no non-animal alternatives. Segregation of animal origin raw materials is desirable in facilities where both animal and non-animal origin components are used. Separation of equipment and manufacturing areas is preferred, but if not possible, validated cleaning procedures should be employed and documented.

Chain of custody of the raw materials is a potential issue as global movement of materials may result in passage through geographic regions or facilities with poor control of pathogens. Traceability of raw materials, including handling and storage is important to document proper source and segregation. Several types of documentation exist to demonstrate proper chain of custody. A media manufacturer should maintain a specification database showing the source, vendor, country of origin, manufacturing process, and grade of each raw material, allowing generation of a

Table 5.4 Excerpt from supplier detail report for DMEM/F12. Highlighted text shows there has been a change in the secondary manufacturing process for L-histidine HCl H₂O from a Japanese supplier

GIBCO™ Invitrogen Corporation		DMEM/F-12 Supplier Detail Report Catalog Number 11320-033	
Component Description	**Animal Derived**	**Process**	**Country of Mfg**
L ALANINE	N	Enzymatic from Bacteria, nonanimal	USA
L ALANINE	N	Enzymatic reaction from L-Aspartic acid, non-animal.	JAPAN
L ALANINE	N	Enzymatic reaction; chemical and plant derived.	Japan
GLYCINE	N	Chemical synthesis, nonanimal	USA
GLYCINE	N	Chemical Synthesis, nonanimal	Japan
L ARGININE HCL	N	Plant fermentation from Glucose	JAPAN
L ASPARAGINE H2O	N	Chemical Synthesis-nonanimal	ITALY
L ASPARAGINE H2O	N	Chemical synthesis, nonanimal	Japan
L ASPARTIC ACID	N	Fermentation from Fumeric Acid-nonanimal	China
L ASPARTIC ACID	N	Chemical purification from crude L Aspartic Acid, nonanimal	China
L CYSTEINE HCL H2O	N	Hybrid process combining organic synthesis with fermentation technology to mfg Cystine. Electronic reduction to convert to Cysteine.	JAPAN
L CYSTINE 2HCL	N	Enzymatic reaction, electric reduction	JAPAN
L CYSTINE 2HCL	N	Synthetically produced Cystine base (Japan) converted to HCl form via further chemical synthesis.	USA
L CYSTINE 2HCL	N	Chemical-synthesis-nonanimal	USA
L GLUTAMIC ACID	N	Plant fermentation-from sugar cane-nonanimal process and raw materials	Brazil
L GLUTAMIC ACID	N	Chemical removal of sodium from MSG, nonanimal	Japan
L HISTIDINE HCL H2O	N	Fermentation from vegtable; nonanimal.	USA
L HISTIDINE HCL H2O	N	Fermentation of sugar-secondary process has been changed from cow milk peptone to soy peptone for lots ending in "0248" and higher	JAPAN

specific report for a given formulation (*Table 5.4*) documenting the source (animal or non-animal), process employed and country of origin for every component.

Specifications are usually linked to a specific compendial reference (USP, EP, JP, etc). In some cases, no compendial reference exists, necessitating assignment of specifications based upon vendor information and/or prior experience with the specific raw material or a related compound. In addition to analytical tests specified in a given compendium, biological performance of the raw materials that are specific for a critical application may be desirable. This is often used for raw materials of an undefined nature like protein hydrolysates or lysates, which tend to exhibit lot-to-lot variability in performance. Key specifications and test results are reported in the Certificate of Analysis for each lot of finished product. It is important that the vendor has a suitable change control policy in effect to document any modifications to specifications. Changes may be vendor-initiated, user-initiated, related to a change in the relevant compendial reference or result from a regular review of the specifications in light of new data. It should be possible for the end user to request notification of specific changes to a particular product. Certificates of Origin are used to document country of origin of the raw material. For nonmilk, ruminant-derived components, a certificate of suitability is often requested. This certificate is issued by the European Directorate for the Quality of Medicines (EDQM) and certifies suitability to a monograph in the European Pharmacopoeia. The dossier submitted to the EDQM contains general information relating to the history of the product and presence of a quality system, as well as specific information about the country, animal and tissue of origin of the raw material, manufacturing process and traceability. An expert report containing a critical evaluation of the content is also included in the dossier. A manufacturer should also have a supplier policy that specifies the level of testing required by the supplier and/or the purchaser. A supplier policy might include the following categories (which are raw material-specific).

- *Provisional.* Requires highest level of testing. Requires Material Review Board approval for manufacturing use.
- *Approved.* Testing required for first receipt of each lot; thereafter, selective tests are performed (e.g. identity).
- *Qualified.* Appearance and identification testing done on first receipt of each lot. Subsequent receipts receive documentation review only.
- *Certified.* Documentation review required for release.

Since a typical media manufacturer may have hundreds of suppliers providing thousands of raw materials, it is not practical to send an audit team to every supplier on a regular basis. One approach to this problem is to have a detailed questionnaire that is filled out by each supplier. Information requested includes:

- origin (plant, animal, chemical, genetically modified organism (GMO), etc.
- details of production process used to manufacture/isolate;
- point in manufacturing process at which animal derivatives are introduced;
- details of steps taken to minimize adventitious agent introduction;
- cleaning validation/elimination of cross-contamination.

Regular updating of supplier information, coupled with on-site audits when appropriate and proactive efforts to identify new sources of non-animal materials serve to minimize risk and maintain credibility. These activities have become even more critical for manufacturers with multiple sites as the ability to move animal-origin materials around the world has become increasingly difficult, requiring extensive documentation of the source and history of the material and sometimes also treatment to reduce risk of transmission of adventitious agents.

5.3.2 Mitigate

When no alternative to an animal-derived component exists, it is sometimes possible to employ strategies that will mitigate the risk of transmission of adventitious agents through the application of various treatment regimes to the raw materials. Of these, gamma irradiation is the most widely used, but thermal (Danner *et al.*, 1999; Plavsic, 2000), UV light (House *et al.*, 1990; Spire *et al.*, 1985) and chemical (Dichtelmuller *et al.*, 1993; Spire *et al.*, 1984) treatment methods have also been described. Key to any mitigation strategy is trend analysis to define the anticipated contaminants, establish safety margins and confidence levels for the proposed treatment and to validate the treatment against a broad-spectrum challenge panel of representative agents in an appropriate scaled-down model system (Jayme and Smith, 2000). Fetal bovine serum and trypsin are two commonly used animal-derived materials that have been successfully used following gamma irradiation. It should be noted that biological performance of the treated material may be altered as a result of the mitigation strategy (Daley *et al.*, 1998). Further, some adventitious agents are resistant to treatments like gamma irradiation (Plavsic and Bolin, 2001). It is important to have validated cleaning methods when equipment and facilities are used for processing both animal and non-animal materials. Cleaning of formulation tanks and equipment after each batch and prior to the start of the next batch should be done with hot (80°C in the circulation system) water that meets the USP requirements for water for injection. Testing for residual total organic carbon (TOC) should be done on surface swabs and residual water samples to validate the cleaning method. Sanitization by hot caustic rinse may provide added assurance of risk mitigation.

5.3.3 Replace

Whenever possible, animal-derived components should be replaced with non-animal origin components of equal specifications. If a facility changes from an animal origin to a non-animal origin raw material, a new raw material stocking number should be used to document the change and to avoid confusion. If the animal origin raw material continues to be used in some processes, separate raw material stocking numbers will minimize the potential for use of the animal origin component in a non-animal process.

Insulin

Insulin has been a component of many serum supplemented and serum-free media over the past 30–40 years. Initially derived from bovine serum for most cell culture applications, recombinant insulin is now commonly used. We performed experiments comparing the effects on hybridoma growth, of two different salt forms of recombinant insulin against a native bovine insulin control (*Figure 5.2*). The sodium recombinant insulin, while exhibiting better solubility characteristics than the zinc form, failed to support the same peak viable cell density as either the zinc recombinant or native bovine forms of insulin. More recent work has confirmed our observations that for most mammalian cell lines, insulin is not necessary if the composition of the medium includes one or more insulin mimetic compounds (Wong *et al.*, 2004). The availability of recombinant insulin (expressed in yeast or *E. coli*) appears to remove it from the list of animal-origin culture media components. However, the current manufacturing process for recombinant insulin involves expression of a longer pro-peptide form of the protein, which is then processed using an animal-derived enzyme to yield the active form. Recombinant insulin would therefore not meet the strictest definition of animal-origin-free because of the use of an animal-derived enzyme in the secondary process.

Transferrin

Human transferrin has been widely used in serum-free cell culture media, but some have chosen to employ transferrin from other mammalian

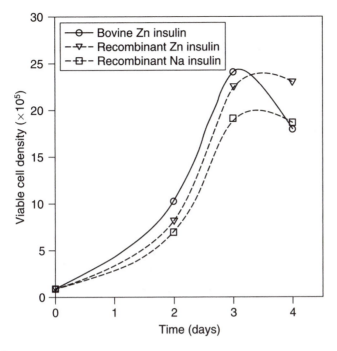

Figure 5.2

Effect of different insulins on hybridoma growth. Recombinant zinc insulin supported similar growth kinetics to native bovine insulin; recombinant sodium insulin did not support as high a peak viable cell density.

species, such as bovine or porcine. Porcine transferrin tends to have activity similar to that of human transferrin for most cultured cell applications, while bovine transferrin has lower activity and must often be used at three to four times the concentration of human transferrin to obtain a similar effect (data not shown). Iron toxicity may occur at high levels of transferrin. Transferrin has been cloned (Uzan, 1984) but must be expressed in mammalian cells for proper glycosylation and is therefore not a cost-effective solution for large-scale culture processes. Pasteurization of animal-derived transferrin removes some risk of transmission of adventitious agents, but newer serum-free formulations have been able to substitute either organic or inorganic iron chelating compounds for transferrin. In our laboratories, we were able to successfully demonstrate the need for iron in serum-free cultures of HEK 293 cells (*Figure 5.3*) by removing the transferrin from a prototype serum-free formulation. When either of two iron chelators (A or B) was substituted for transferrin, growth and peak viable cell densities were comparable to results obtained with human transferrin. It should be noted that the optimal concentrations for A and B were previously determined in other experiments and were not the same. This highlights the need for titering any new medium component.

We compared growth over serial passages of HEK 293 cells and PER.C6 cells in media containing different iron chelators (*Table 5.5*). Not all iron chelators supported growth of these cells and some that initially supported

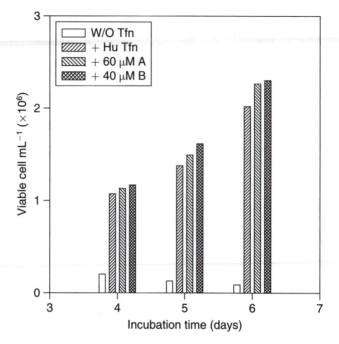

Figure 5.3

Growth of HEK 293 cells in serum-free media with various iron carriers. Iron chelator A or B supported similar growth kinetics and peak cell densities compared to human transferrin.

Table 5.5 Growth of HEK 293 and PER.C6® cells in serum-free media with various iron carriers

Iron chelator	Performance relative to transferrin
Ferric citrate complex	+++
Ferrous sulfate EDTA	++
Ferric sulfate EDTA	+
2-Hydroxypyridine N-oxide	+++
N-hydroxysuccinamide	0
Glycine hydroxyamate	0
Sorbitol ferric chloride	+++/0
Sodium nitroprusside	0
Ferridoxin	0
Tropolone	+++/0
Desferroxamine mesylate ferric chloride	0

+++ = Equivalent to transferrin; 0 = not suitable replacement; +++/0 = performance deteriorates with time.

growth to a level equivalent to that observed with transferrin, later failed to maintain an acceptable level of performance. There are many considerations for the use of iron carriers. For example, there are iron carriers that are complexes of the iron with other molecules (e.g. ferrous sulfate EDTA) and there are uncomplexed forms (e.g. ferrous sulfate). The affinity of the carrier

Table 5.6 Recovery of IgG from an MEP HyperCel column

	Fe (% recovery)	IgG (% recovery)
(A) Low recovery: untreated culture supernatant		
Feed stock	100	100
Flow-through	26.9	9.4
Wash	12.5	6.0
Elution	4.6	40.6
(B) Improved recovery: 10 mM tetrasodium EDTA treated		
Feed stock	100	100
Flow-through	93.3	10.3
Wash	ND	10.3
Elution	ND	82.5

for iron is important; the carrier should be able to transport iron to the cells, but should also be able to release the iron to the cells. Some chelators may bind the iron so tightly that it may be unusable by the cells. Similarly, if the chelator is not specific for iron (e.g. EDTA), other species (e.g. other cations) may compete with iron for the chelator. Solubility of the iron carrier may be a problem and can result in clogging of filtration membranes. The carrier may have only limited stability in aqueous solution and may cause toxic effects, either directly or indirectly through release of iron to react with oxygen species, or through binding of other components, making them unavailable to the cells. Intellectual property restrictions exist with some iron carriers (e.g. tropolone; Metcalfe *et al.*, 1994). Finally, iron carriers may interfere with detection and purification of expressed proteins. We have encountered iron carriers that absorb light at 280 nm, interfering with spectrophotometric estimation of protein content of culture supernatants and column eluents. *Table 5.6* shows iron interference with a hydrophobic charge interaction chromatographic (HCIC) method for purifying IgG. *Table 5.6A* shows the low recovery of IgG after untreated culture supernatant was passed through a column of MEP HyperCel. Recovery of IgG was doubled by pretreatment of the culture supernatant with 10 mM tetrasodium EDTA (*Table 5.6B*).

Plant hydrolysates

Plant hydrolysates have become increasingly popular as media additives to improve growth, productivity or both (Lobo-Alfonso *et al.*, 2005). The useful range in mammalian cell culture media tends to be between 0.1 and 5.0 g L^{-1}, although lower or higher amounts have been reported. As for other additives to cell culture media, it is important to titrate to find the concentration that provides the optimal biological effect. There are many plant-derived hydrolysates commercially available, originally developed for the food flavor industry and as components of microbial fermentation broths. Applications for hydrolysates in these industries far exceed their use in mammalian cell culture. This has caused concern for manufacturers of human therapeutics who use plant hydrolysates made by processes that were never designed for cGMP mammalian cell culture. There are many different plant sources used for hydrolysates, and manufacturing processes can vary substantially between the different hydrolysate suppliers. Raw materials may have been sourced from multiple geographies, grown in a variety of soil conditions, exposed to pesticides and herbicides and may have come in contact with animals or animal by-products. It is therefore important to evaluate different plant sources, manufacturing processes and manufacturing facilities before deciding on employing a plant hydrolysate in a bioproduction manufacturing process. Lot-to-lot variability can be significant, necessitating evaluation of multiple lot samples, and ideally, implementation of raw material specifications that include biological performance demonstrated in a relevant cell line. Hydrolysate manufacturing facilities and processes should be audited to ensure either separation of plant from animal processing (including the use of non-animal enzymes for processing of the plant material), or if the facility is dual use, validated cleaning procedures should be employed. There appears to be no universal hydrolysate for all cell types and it has become popular in recent years to mix different hydrolysate to obtain the desired improvement in growth and/or expression of recombinant product. Ultrafiltration of the hydrolysate may be done to remove poorly soluble high molecular weight species with the added benefit of a reduction in endotoxin load. In general, hydrolysates tend to exhibit good stability if stored properly, are made in large lot sizes with a cost that is low relative to more defined components. They have been shown to boost cell growth and/or productivity in many systems, but they have also been demonstrated to interfere with upstream

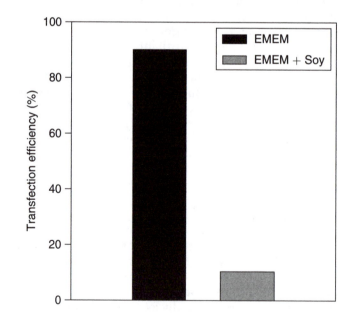

Figure 5.4

Soy hydrolysate-mediated interference with cationic lipid transfection in CHO-K1 cells. CHO cells grown in a serum-free medium were transfected with lipofectamine to express beta galactosidase. DNA/lipofectamine complexes were delivered to the cells in either serumless EMEM or serumless EMEM + soy. Reduced transfection efficiency was observed in the presence of soy.

processes like transfection (*Figure 5.4*) and downstream processes like colori-
metric endotoxin determination and purification methods.

Lipids

Lipids have many roles in cells including membrane structural components, storage and transport of metabolic fuels, oligosaccharide transport,
cell recognition, species specificity, tissue immunity and signal transduction. Many cultured cells are able to make lipids required for metabolic
processes from acetyl CoA. However, many NS0 and related cells are sterol
auxotrophs, lacking a key enzyme in the cholesterol biosynthetic pathway
(Sato *et al.*, 1988). These cells require supplementation of culture media
with a source of sterol. Animal sera are able to provide sterols and other
lipids, but for serum-free systems, other options can include:

- serum fractions (lipoproteins like EX-CYTE®);
- chromatographically purified serum albumin (AlbuMAX®);
- lipids complexed to albumin;
- lipids dissolved in ethanol;
- PLURONIC®-based lipid emulsions;
- Cyclodextrin-based lipid complexes.

Requirements for animal origin-free components in many processes have
reduced the use of serum fractions and serum-derived proteins. Lipids
dissolved in ethanol can be added to aqueous media, but generally must be
added in amounts less than 1 mL L^{-1} (v/v). PLURONIC®-based lipid supplements are prone to cracking of the emulsion and may not provide sufficient
levels of sterol before the PLURONIC® becomes toxic to the cells cultured in
serum-free media. PLURONIC® has also proved problematic in downstream
purification protocols. Cyclodextrins are ring-like sugar molecules with the
ability to encapsulate other compounds in a hydrophobic inner core. Many
options exist for choice of cyclodextrin (alpha, beta or gamma) and specific
modifications (e.g. methyl beta cyclodextrin, hydroxypropyl beta cyclodextrin, etc). The interaction between the cyclodextrin molecule host and its
guest molecule is dynamic, so care must be taken to use the right ratio of
cyclodextrin to lipid to ensure solubilization in aqueous media. Too much
cyclodextrin can be toxic, while too little may result in poor solubility of
lipids and removal by filtration media. For this reason, cyclodextrin-lipid
complexes may require aseptic addition to culture media in some instances.
We have developed several lipid supplements utilizing cyclodextrin as the
carrier. The sterol originally used in these supplements was ovine cholesterol derived from wool grease. While the risk of transmission of adventitious agents from ovine cholesterol is most likely minimal owing to the
geographic source of the wool and the hot caustic treatment used to extract
the sterol, the fact that it is of animal origin prompted the search for a non-
animal replacement. Several plant sterols were tested, but as shown by the
results of a representative experiment in *Figure 5.5*, biological performance
was not equal to that obtained using ovine cholesterol. A synthetic cholesterol, made by chemical modification of a plant-derived starting material
was tested against the ovine cholesterol with the results shown in *Figure
5.6A and B.*

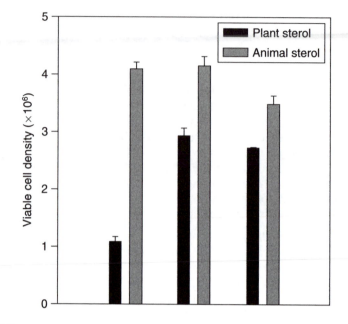

Figure 5.5

Wild-type NS0 cells were cultured in a protein-free, chemically defined medium to which was added a cyclodextrin-lipid complex consisting of methyl beta cyclodextrin and either plant sterol or animal sterol. The plant sterol did not support cell growth to the same level as the animal sterol over 11 passages.

Dissociating enzymes

The most commonly used enzyme for cell culture applications is porcine pancreatic trypsin. The animal origin of this reagent and variability in performance between different lots is cause for concern for use in processes to manufacture human biotherapeutics. The presence of other enzymes like chymotrypsin or pancreatin may contribute to variability in performance. Stability of porcine trypsin is poor unless stored frozen. Porcine trypsin is often irradiated to minimize risk of transmission of porcine parvovirus and other adventitious agents, but this treatment does nothing to reduce the potential for performance variability and poor stability. Alternatives to porcine trypsin include EDTA (Versene), bacterial-derived enzymes (Collagenase, Dispase), transgenic corn-derived protease and fungal-derived recombinant protease. We have successfully developed the recombinant fungal protease for cell culture dissociation applications. This material is highly purified and lacks the contaminants commonly found in porcine trypsin (*Figure 5.7*). The recombinant protease (TrypLE™ Express and TrypLE™ Select) has high levels of activity and stability maintaining enzymatic activity following storage at 4°C weeks longer than porcine trypsin (*Figure 5.8*). TrypLE™ has been successfully used for dissociation of a variety of cell lines (Nestler *et al.*, 2004).

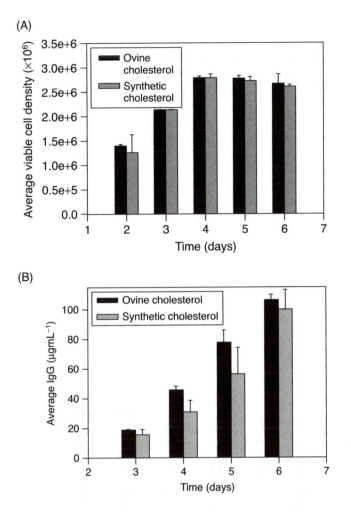

Figure 5.6

Recombinant NS0 cells were cultured in a protein-free, chemically defined medium to which was added a cyclodextrin-lipid complex consisting of methyl beta cyclodextrin and either synthetic cholesterol or animal sterol. The synthetic cholesterol supported cell growth (A) and IgG expression (B) to equivalent levels as the ovine cholesterol.

5.4 Summary and considerations for the future

The definition of what constitutes animal-origin material is evolving to include secondary and tertiary processing steps. As a result of this change in definition, there has been a change in requirements for traceability and documentation for each medium component and reagent. Until such time as all raw materials and reagents have been declared totally animal origin free (to whatever level of processing required), there will be a call to separate equipment and facilities used to process animal origin materials

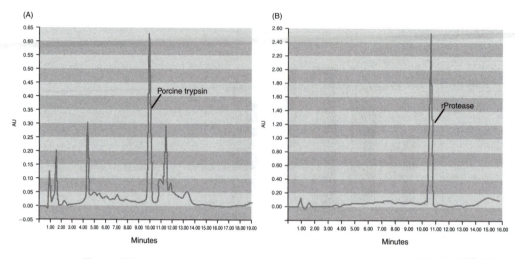

Figure 5.7

Protease purity. HPLC chromatograph A illustrates a typical Hamilton PRP-3 column separation of porcine trypsin. Note the multiple peaks indicating impurities. HPLC chromatograph B demonstrates the purity of the recombinant protease sample.

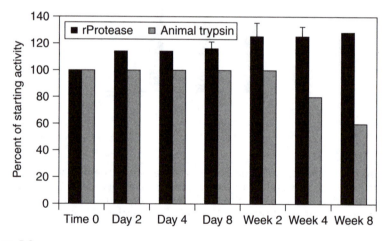

Figure 5.8

Stability of 1X recombinant protease and 0.25% animal trypsin after storage at 4°C. Recombinant protease maintained activity weeks longer than porcine trypsin. Values obtained from Ac-Arg-pNA kinetic enzyme absorbance assay and expressed as percent of starting activity at time 0. Error bars are ±1 standard deviation.

from those used for non-animal origin materials. If this is not possible, validated cleaning procedures will need to be demonstrated. The cleaning validation may need to be periodically updated as new adventitious agents are identified as being potential risks. Similarly, risk mitigation strategies like gamma irradiation will need to be re-evaluated and revalidated as new adventitious agents are described. Replacement of animal origin components with non-animal alternatives represents an attractive approach to address safety concerns. However, a major consideration is in how non-animal origin raw materials are qualified for use in the production of biotherapeutics and bioprophylactics. Some replacement components are chemically identical to the animal origin components and would therefore not be expected to require extensive validation. However, the source materials and processes used to manufacture the non-animal replacement component may be quite different from the animal component and there may be trace contaminants found in one, but not in the other. Such may be the case for plant derived materials, which may be theoretically free from animal associated adventitious agents, but may in fact have been exposed to animals while the plants were being grown in the field and during processing by human operators. Similarly, plants may have been exposed to herbicides, pesticides and/or potentially toxic substances present in the soil, air or water. Contamination of plant-based material with fungi may result in the presence of mycotoxins (FDA report, 2004) which can affect cultured cells (Babich and Borenfreund, 1991; Hsia *et al.*, 1986) and in humans can have variable effects (Peraica *et al.*, 1999). Often considered as contaminants in food (FDA monograph), these potentially toxic metabolites may precipitate a re-evaluation of the growing use of plant-based raw materials in cell culture processes. A higher level of testing for critical potential contaminants of plant-derived material may eventually be required. Analytical fingerprinting may become a criterion for acceptance of incoming raw materials. In some cases where a replacement component has the same structure as the animal origin component and in cases where an animal origin component has a different structure than the replacement component, validation employing biological assay systems may be appropriate. Such testing was done for both non-animal cholesterol and for the recombinant protease used to replace porcine trypsin. In the case of cholesterol, the identical chemical structure correlated well with equivalent biological performance. In the case of recombinant protease, the higher activity relative to porcine trypsin prompted an adjustment in concentration of the recombinant enzyme to achieve similar biological effects. Titration of replacement components is an important consideration since cell culture components often exhibit an optimal concentration range, below or above which, the effect may be neutral or negative. The potential for non-animal origin components to complicate upstream and downstream process was demonstrated by the effect of soy hydrolysate on cationic lipid-mediated transfection and by the effect of an iron chelator on IgG purification recovery. Recognition of the potential for such phenomena to occur will save time, effort and cost in developing the overall process in which cell culture is just one part.

There are clearly many options to customize cell culture media and processes to address concerns about raw material availability, performance and safety. A proactive, holistic approach in which biomanufacturers partner with suppliers to exert tight control over raw material sourcing and quality at all stages of the process will benefit the ultimate end user, the patient.

Acknowledgments

The author wishes to acknowledge the assistance of the following people who contributed data and other information to this chapter: Susan Aube, Caroline Briggs, Susan Cameron, Delia Fernandez, Melanie Kester, Cynthia Louth, David Jayme, David Judd, Robert Keem, Lori Nestler, Leslie Madigan, Jim Mecca, Navin Patel, William Paul, Paul Price, Morven Strachan, Brad Stiles, Kelli Tanzella, Lia Tescione, Mary Lynn Tilkins, and Jennifer Walowitz.

References

Babich H and Borenfreund E (1991) Cytotoxicity of T-2 toxin and its metabolites determined with the neutral red cell viability assay. *Appl Environ Microbiol* **57**: 2101–2103.

Barnes D and Sato G (1980) Methods for growth of cultured cells in serum-free medium. *Anal Biochem* **102**: 55–270.

Daley JP, Danner DJ, Weppner DJ and Plavsic ZM (1998) Virus inactivation by gamma irradiation of fetal bovine serum. *FOCUS* **20**: 86–88.

Danner DJ, Smith J and Plavsicz M (1999) Inactivation of viruses and mycoplasmas in fetal bovine serum using 56°C heat. *BioPharm* **12**: 50–52.

Dichtelmuller H, Rudnick D, Breuer B and Ganshirt KH (1993) Validation of virus inactivation and removal for the manufacturing procedure of two immunoglobulins and a 5% serum protein solution treated with beta-propiolactone. *Biologicals* **21**: 259.

Eagle H (1955) Nutrition needs of mammalian cells in tissue culture. *Science* **122**: 501–504.

EMEA/410/01 rev 2 (October 2003). http://www.emea.eu.int/pdfs/human/bwp/TSE%20NFG%20410-rev2.pdf Note for guidance on minimising the risk of transmitting animal spongiform encephalopathy agents via human and veterinary medicinal products.

FDA report (2004): http://www.cfsan.fda.gov/~comm/cp07002.html Mycotoxins in Imported Foods (FY 02/03/04) Chapter 7 – *Molecular Biology and Natural Toxins*.

Gruber DF and Jayme DW (1994) Cell and Tissue Culture Media: History and Terminology, in Cell Biology: A Laboratory Handbook, vol. 3. Academic Press Inc., New York, pp. 451–458.

Ham RG (1965) Clonal growth of mammalian cells in a chemically defined, synthetic medium. *Proc Natl Acad Sci USA* **53**:288–293.

House C, House JA and Yedloutschnig RJ (1990) Inactivation of viral agents in bovine serum by gamma irradiation. *Can J Microbiol* **36**: 737.

Hsia CC, Gao Y, Wu JL and Tzian BL (1986) Induction of chromosome aberrations by fusarium T-2 toxin in cultured human peripheral blood lymphocytes and Chinese hamster fibroblasts. *J Cell Physiol Suppl* **4**: 65–72.

Jayme D, Watanabe T and Shimada T. (1997). Basal medium development for serum-free culture: a historical perspective. *Cytotechnology* **23**: 95–101.

Jayme DW (1999) An animal origin perspective of common constituents of serum-free medium formulations. In: *Animal Sera, Animal Sera Derivatives and Substitutes Used in the Manufacture of Pharmaceuticals: Viral Safety and Regulatory Aspects. Developments in Biological Standardization 99,* (eds F. Brown, T. Cartwright, F. Horaud and J.M. Speiser). Karger, Switzerland, pp. 181–187.

Jayme DW and Blackman KE (1985) Review of culture media for propagation of mammalian cells, viruses and other biologicals. In: *Advances in Biotechnological Processes*, vol. 5 (eds A. Mizrahi and A.L. van Wezel). Liss, NY, pp. 1–30.

Jayme DW and Gruber DF (1994) Development of serum-free media and methods for optimization of nutrient composition. In: *Cell Biology: a Laboratory Handbook*, vol 1. Academic Press Inc., New York, pp. 18–24.

Jayme DW and Smith SR (2000) Media formulation options and manufacturing process controls to safeguard against introduction of animal origin contaminants in animal cell culture. *Cytotechnology* **33**: 27–36.

Lobo-Alfonso J, Price P and Jayme D (2005) Benefits and limitations of protein hydrolysates as components of serum-free media for animal cell culture applications. In: *Protein Hydrolysates in Nutrition and Biotechnology*, (ed. Vijay K. Pasupuleti). Kluwer Academic Publishers (forthcoming).

Merten OW (1999) Safety issues of animal products used in serum-free medium. In: *Animal Sera, Animal Sera Derivatives and Substitutes used in the manufacture of Pharmaceuticals: Viral Safety and Regulatory Aspects. Developments in Biological Standardization* 99, (eds F. Brown, T. Cartwright, F. Horaud and J.M. Speiser). Karger, Switzerland, pp. 167–180.

Metcalfe H, Field RP and Froud SJ (1994) The use of 2-hydroxy-2,4,6-cycloheptarin-1-one (tropolone) as a replacement for transferrin. In: *Animal Cell Technology, Products of Today, Prospects for Tomorrow* (eds R.G. Spier, J.B. Griffiths and B. Moignier). Butterworth-Heinemann, pp. 88–90.

Moore GE, Gerner R and Franklin H (1967) Culture of normal human lymphocytes. *J Am Med Assoc* **199**: 519–524.

Nestler L, Evege E, McLaughlin J, Munroe D, Tan T, Wagner K and Stiles B (2004). TrypLE express: A temperature stable replacement for animal trypsin in cell dissociation applications. *Quest* **1**: 42–47.

Peraica M, Radic B, Lucic A and Pavlovic M (1999) Toxic effects of mycotoxins in humans. *Bull World Health Organ* **77**: 754–766.

Plavsic ZM (2000) Effect of Heat treatment on four viruses inoculated into BSA and bovine transferrin solution. *BioPharm* **13**: 54–56.

Plavsic ZM and Bolin S (2001) Resistance of porcine circovirus to gamma irradiation. *BioPharm* April, 2001: 32–36.

Sato D J, Cao H-T, Kayada Y, Cabot MC, Sato HH, Okamoto T and Welsh CJ (1988) Effects of proximate cholesterol precursors and steroid hormones on mouse myeloma growth in serum-free medium. *In Vitro Cell Dev Biol* **24**: 1223–1228.

Spire B, Barre-Sinoussi F, Montagnier L and Chermann JC (1984) Inactivation of lymphadenopathy associated virus by chemical disinfectants. *Lancet* **20**: 899–901.

Spire B, Dormont D, Barre-Sinoussi F, Montagnier L and Chermann JC (1985) Inactivation of lymphadenopathy-associated virus by heat, gamma rays, and ultraviolet light. *Lancet* **26**: 188–189.

Uzan G, Frain M, Park I, Besmond C, Maessen G, Trepat JS, Zakin MM and Kahn A. (1984) Molecular cloning and sequence analysis of cDNA for human transferrin. *Biochem Biophys Res Commun* **119**: 273–281.

Wong VT, Ho KW and Yap MGS (2004) Evaluation of insulin-mimetic trace metals as insulin replacements in mammalian cell cultures. *Cytotechnology* **45**: 107–115.

Glycosylated proteins

Post-translational modifications of recombinant antibody proteins

6

Roy Jefferis

6.1 Introduction

Following the success of the human genome project, the transcriptome, proteome, glycoproteome, glycome etc. have become foci of interest (Mann and Jenson, 2003). An understanding of human complexity is being sought not in the total number of genes but in the protein products of those genes; and other molecules synthesized through the action of protein products. The transcriptome/proteome exceeds the genome due to differential splicing of nuclear RNA, protein post-translational modifications (PTMs) and protein products generated and/or released in cascade reactions (e.g. coagulation, complement activation etc.). It has been estimated that human identity/integrity depends on the action of 10^6 individual molecules (http://us.expasy.org/sprot/hpi/; Bauman and Meri, 2004; O'Donovan *et al.*, 2001). Possibly, the most frequent and diverse PTM is glycosylation since it is estimated that ~50% of genes encode for proteins with the potential to bear N-linked glycans, i.e. they express the Asn-X-Ser/Thr motif, where X may be any amino acid except proline (Wong, 2005). Additionally, O-linked glycans add to glycoprotein complexity; however, their potential presence cannot be predicted from gene or protein sequences. Defects in genes contributing to N- and O-linked glycosylation pathways result in congenital disorders of glycosylation (CDG) having serious medical consequences (Butler *et al.*, 2003; Freeze, 2002). Changes in the glycosylation profiles of specific proteins may serve as disease markers (Axford *et al.*, 2003; Gu *et al.*, 1994; Holland *et al.*, 2002, 2006; Ito *et al.*, 1993; Parekh *et al.*, 1985; Takahashi *et al.*, 1987; Youings *et al.*, 1996) whilst the significance of other disease related changes is yet to be elucidated (Poland *et al.*, 2005).

It will be evident that a recombinant protein should, ideally, exhibit the same PTMs as the endogenous protein product. However, it is important to recognize that the structure determined for an endogenous protein is that of a molecule that has had a residence time in a body compartment/fluid prior to being subject to multiple isolation and purification protocols. The structure of this purified product could differ from that of the nascent molecule secreted from its tissue of origin. Similarly, recombinant proteins are synthesized in an 'alien' tissue (CHO, NS0 cells etc.), are exposed to the culture medium, products of the host cell line and subject to rigorous downstream and formulation processes. Lack of structural fidelity can impact on function, stability and immunogenicity; an immune response may impact therapeutic efficacy and/or result in harmful reactions (side-effects) (Sinclair and Elliott, 2005; Smalling et al., 2004). Glycosylation and other PTMs have been shown to be species, tissue and gender specific (Davies et al., 2001; Gomord et al., 2005; Hadley et al., 1995; Jefferis, 2005; Raju et al., 2000; Shields et al., 2002; Shinkawa et al., 2003; Umana et al., 1999; van den Nieuwenhof et al., 2000). Currently, protein therapeutics have a shelf life of 18–24 months, which is testament to a lack of structural integrity or suboptimal formulation.

Prokaryotic systems (e.g. E. coli) are unable to produce glycosylated proteins and the aglycosylated protein product may form an inclusion body that has to be extracted, solublized, and refolded in vitro. In contrast, yeast systems add very high mannose structures whilst insect cells add pauci-mannose structures. Plants may differentially glycosylate proteins but consistently add α-(1-3) fucose and β-(1-2) xylose sugars that are reported to be immunogenic/allergenic in humans (Gomord et al., 2005). Cellular productivity and PTMs are influenced by cell culture conditions, for example temperature, growth rate, media composition (Andersen et al., 2000; Mimura et al., 2001a; Rodriguez et al., 2005), and the addition of butyrate etc. has been shown to increase production and influence the glycoform profile of glycoprotein products (Mimura et al., 2001a). Considerable success has been reported for increased productivity of antibody in CHO cell lines, with levels of 5 g L^{-1} being achieved and 10 g L^{-1} being set as a goal (Birch, 2005). However, high production levels may overwhelm the PTM machinery resulting in poor product quality; it is essential, therefore, to characterize a product at an early stage in clone selection to optimize both productivity and quality. Essential nutrients may also compromise product quality, e.g. glycation through the nonenzymatic addition of glucose (Harris, 2005; Lapolla, 2001) or oxidation of methionine side chains (Harris, 2005). The mammalian CHO, NS0 and Sp2/0 cell lines produce an endogenous carboxypeptidase-b that differentially cleaves the C-terminal lysine residues from antibody heavy chains, adding structural and charge heterogeneity (Harris, 2005). Thus, CHO cells may be particularly inappropriate for the production of the complement proteins C3a, C4a etc. that bear functionally significant C-terminal arginine residues; the desArg forms of these proteins having a different profile of activities.

6.2 Common post-translational modifications

Proteins can display an extraordinary range of PTMs (>340) (Walsh and Jefferis, 2006; http:// us.expasy.org/sprot/hpi/). The recombinant protein therapeutics currently licensed are either soluble proteins, normally secreted from the cells in which they are biosynthesized, or soluble extracellular domains of membrane proteins. The normal proteolytic processing of pre- and pro-proteins, e.g. removal of N-terminal methionine, signal peptide and fidelity of disulfide bond formation has not presented a problem when produced in mammalian cells. The PTM that has been a major focus of interest has been glycosylation. Whilst it is estimated that ~60% of eukaryotic proteins are phosphorylated and acetylation, these PTMs are mostly confined to proteins mediating intracellular signalling, trafficking and control functions. Such molecules have not been developed as recombinant protein therapeutics, to-date.

6.3 Recombinant antibody therapeutics

The last two sentences of the ground-breaking Kohler and Milstein paper read: 'Such cells can be grown *in vitro* in massive cultures to provide specific antibody. Such cultures could be valuable for medical and industrial use.' (Kohler and Milstein, 1975). Recombinant antibody therapeutics (rMAbs) are predicted to become the largest family of disease-modifying drugs available to clinicians (Tanner, 2005) and the 2006 market value for therapeutic rMAb alone is predicted at $15 billion! Their efficacy results from specificity for a target antigen and biological activities (effector functions) activated by the immune complexes formed. Eighteen rMAbs are currently licensed and hundreds are in clinical trials or under development. The biopharmaceutical industry has met the challenge to produce rMAbs, although productivity, cost, and potency remain to be optimized. All antibody therapeutics currently licensed are produced by mammalian cell culture, utilizing Chinese hamster ovary (CHO), mouse NSO or mouse Sp2/0 cell lines; other systems under development and evaluation, include transgenic animals, yeasts, fungi, plants etc. All rMAbs have shown a potential for immunogenicity whether presented as mouse, chimeric, humanized or fully human sequences. These responses are referred to as human anti-mouse antibody (HAMA), human anti-chimeric antibody (HACA) or human anti-human antibody (HAHA) (Mirik *et al.*, 2004); the promise that fully human antibodies may not be immunogenic has not been realized for Humira (Adalimumab), generated by phage display from a human heavy- and light-chain library, since 12% of patients have been shown to produce anti-Humira antibodies. Such antibody responses will prejudice treatment if they are neutralizing, lead to clearance of the therapeutic or sensitize the patient for severe reactions on re-exposure (Smalling *et al.*, 2004).

The effectiveness of rMAb in oncology depends on sensitizing target cells for subsequent killing by the mechanisms of antibody dependent cellular cytotoxicity (ADCC) and/or complement dependent cytotoxicity (CDC). It is unequivocally established that these effector functions are dependent on appropriate glycosylation of the rMAb (Jefferis, 2005). Glycosylation has been a focus of interest for the biopharmaceutical industry for the past several years and cell lines have been engineered to optimize the product for ADCC and CDC; for example, by the addition of galactose, bisecting *N*-acetylglucosamine or exclusion of the addition of fucose (see below). The FDA requires that a consistent human-type glycosylation be maintained for rMAbs, irrespective of the system in which they are produced. This presents a continuing challenge as the biopharmaceutical industry continues to develop cell lines and culture conditions for increased productivity (5–10 g L^{-1}). Proteomic studies of the producer cell lines are being initiated in attempts to identify parameters influencing production levels and product quality (Alete *et al.*, 2005; Mann and Jenson, 2003). To date, attention has focused on glycosylation at a single site (Asn 297). However, ~20% of normal human antibodies have an additional glycosylation site, as does one licensed therapeutic antibody (Erbitux) (Jefferis, 2005).

6.4 Structural and functional characteristics of human antibodies

The original monoclonal antibody technology provided for the production of mouse antibodies to virtually any antigen and was rapidly developed to generate research and diagnostic reagents that have made an invaluable contribution to the discovery and characterization of the human proteome and, indirectly the genome. However, it was evident that mouse antibodies could only have a very limited role as therapeutics in humans, if any, since an ensuing immune response to the foreign mouse protein would, at best, neutralize its effectiveness and, at worse, provoke serious immune mediated reactions. This problem was partially resolved with developments in genetic engineering techniques that allowed the fusion of mouse and human antibody genes with the production of chimeric mouse/human antibodies. These antibodies are comprised of the mouse variable regions, that determine antigen binding specificity, and human constant regions that determine biological effector mechanisms activated by immune complexes.

These developments required insight into the mode of action of antibodies since five classes of antibody (IgG, IgA, IgM, IgE, IgD) have been defined in humans and each has a unique role in providing immune (humoral) protection. All licensed antibody therapeutics, to date, have been of the IgG class. Antibody of the IgG class predominates in human blood, equilibrates with extravascular fluid and activates a wide range of effector functions resulting in the killing, removal, and destruction of pathogens. Further, informed choice must be exercised, however, since in humans there are four subclasses of IgG (IgG1, IgG2, IgG3, and IgG4) that, although of high sequence homology, differ in their abilities to activate downstream biologic functions. Initially, only IgG1 antibodies were developed but, recently, two IgG4 antibodies have been licensed and IgG2 antibodies are in development. These important decisions are predicated on the presumed biologic activities required *in vivo* and the IgG subclass that may be optimal for delivery (Jefferis, 2005; Jefferis *et al.*, 1998; Nezlin and Ghetie, 2004; Woof and Burton 2004).

Whilst the IgG antibody class provides systemic immune protection it is confined to vascular and extravascular fluids within the body. However, we are most prone to infection by pathogens at mucosal surfaces: the respiratory, gastrointestinal, and urinogenital tracts. These exposed sites present live tissue, bathed in fluid and at body temperature to the external environment, conditions ideal for colonization and proliferation of microorganisms. A local or mucosal immune system has evolved that provides for the production of IgA antibody in a form, secretory IgA, that passages to the external mucosal surface by transcytosis (Woof and Mestecky, 2005). There is significant activity aimed at the production of secretory IgA antibodies as prophylactics and therapeutics for topical and oral administration.

6.5 The human IgG subclasses: Options for antibody therapeutics

The human IgG subclasses are enumerated according to their relative concentrations in normal serum. Thus, IgG1, IgG2, IgG3 and IgG4 account for ~60%, 25%, 10%, and 5%, respectively, and is reflected in the frequency of monoclonal IgG proteins in multiple myeloma. The predominance and accessibility of IgG1 subclass proteins and antibodies has allowed for a comprehensive study of their structure and structure/function relationships. Consequently, the IgG1 subclass appeared to be the natural choice in development of antibody therapeutics. This assumption is now under review as it is appreciated that the optimal mode of action, *in vivo*, will vary according to the disease indication.

Normal serum IgG is comprised of thousands (tens of thousands!) of structurally unique antibodies each specific for a unique epitope (antigenic determinant). Given the relative proportions of each IgG subclass present in normal serum it is appropriate to ask whether a specific antibody population will be comprised of similar proportions of each subclass. This is not the case and the broad generalization can be made that for protein antigens IgG1 and IgG3 predominate, whereas for carbohydrate antigens, IgG2 and IgG4 may predominate as a result of chronic antigen stimulation, *Table 6.1* (http://www.researchd.com/rdikits/rdisubbk.htm; Jefferis *et al.*, 1998).

We attempt to rationalize the significance of this 'partition' of responses with respect to the known IgG-Fc functional activities of each IgG subclass, summarized in *Table 6.2*, although it must be emphasized that these assignments represent a gross generalization extrapolated from many *in vitro* studies, often employing heterologous systems (e.g. guinea pig complement). We face a considerable challenge to devise means of determining the effector functions activated *in vivo* when a given rMAb therapeutic is delivered to an individual patient for a given disease indication. Two broad categories of action may be distinguished; (i) the binding of antibody to

Table 6.1 The ligand binding/activation properties of the human IgG

Isotype	IgG$_1$	IgG$_2$	IgG$_3$	IgG$_4$
C1‡	++	−	+++	−
FcγRI	+++	−	+++	++
FcγRII	+	±* +	?	
FcγRIIIa/b	+	−	+	±*†
FcRn	+	+	+	+

‡Immune complexes may also activate the alternate pathway.
*Dependent on FcγR polymorphisms.
†Dependent on glycoform.

Table 6.2 IgG subclass profile of specific antibody responses

Antigen	IgG$_1$	IgG$_2$	IgG$_3$	IgG$_4$
Tetanus toxoid	+++	+	++	+
Polysaccharides	++	+++	+	+
Rhesus-D	+++	−	++	−
Factor VIII	−	−	−	+++
Erythropotein	+	−	−	+++
Phospholipase A$_2$	+++	+	+	+
Phospholipase A$_2$*	+	+	+	+++

*Responses in bee keepers.

cells to sensitize them for immune killing and removal (e.g. in oncology); (ii) binding and neutralization of soluble molecules (e.g. cytokines, toxins in chronic diseases). Multiple parameters may impact on the outcome (Voice and Lachmann, 1997).

The IgG1 and IgG3 subclasses are essentially equivalent in their abilities to activate each of the leucocyte Fc receptors and the classical complement cascade. However, they have different catabolic rates, with serum half-lives of ~21 and ~7 days, respectively. Since IgG is protected from degradation (catabolism) by interaction with the FcRn receptor it may be presumed the affinity of IgG3 for FcRn is lower than that of IgG1. However, IgG1 and IgG3 appear to cross the placenta, an FcRn mediated transcytosis event, with equal facility. The hinge region of the IgG subclasses differs in sequence and length and is a prime site for enzymatic cleavage. The extended hinge region of IgG3 is particularly susceptible and this presents difficulties for its isolation, purification, and stabilization. Consequently, it has not been the isotype of choice in the development of rMAb; except for anti-rhesus-D antibodies for which a prophylactic is being developed containing IgG1 and IgG3 rMAbs at a ratio of 3:1, as observed in patient immune responses (Kumpel, 2002).

It is recognized that the activation of inflammatory cascades may not be required, or may be detrimental, for some antibody based therapies. This has resulted in contemplation of the respective merits of the IgG2 and IgG4 subclasses. Two IgG4 therapeutics have been developed, presumably selected for reduced effector activities, and received regulatory approval. This presumption is open for re-evaluation since it has been demonstrated that IgG4 can activate FcγRI (Jefferis *et al.*, 1998) and FcγRIIIa, depending on the glycoform of the antibody and the allotype of the receptor (Niwa *et al.*, 2005). In contrast IgG2 has been shown only to activate one allotypic variant of the FcγRIIa receptor. Neither IgG2 nor IgG4 have been convincingly shown to activate the classical complement pathway when only human complement proteins are employed (http://www.researchd.com/rdikits/rdisubbk.htm; Woof and Burton, 2004).

6.6 The structure of human IgG antibodies

In its simplest form an individual IgG molecule is composed of two identical light chains and two identical heavy chains comprised of repeating structural motifs of ~110 amino acid residues – referred to as a homology region. A light chain is comprised of two homology regions or domains and an IgG heavy chain of four *Figure 6.1A*. The structural basis of antibody specificity is the unique primary amino acid sequence of the first homology regions of each of the heavy (V_H) and light chains (V_L). The other light and heavy chain domains are referred to as the 'constant domains' since they have a conserved sequence characteristic for a light chain type or heavy chain class or subclass. The tertiary structure of each homology region exhibits a characteristic anti-parallel β-pleated sheet structure that confers stability and function; referred to as the immunoglobulin fold – a ubiquitous structure that defines the members of the immunoglobulin superfamily (Deisenhofer, 1981; Nezlin and Ghetie, 2004; Woof and Burton, 2004). Domains of the light and heavy chains pair in covalent and noncovalent association to form

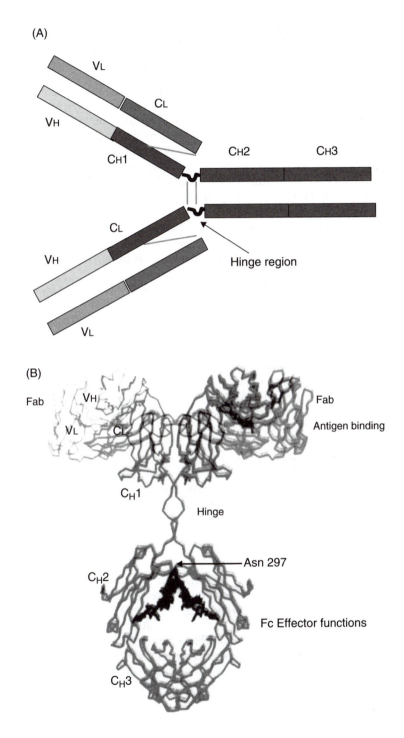

Figure 6.1

(A) The basic four chain structure of the IgG molecule inter H-H and H-L disulfide bridges. (B) α-Carbon backbone structure of the IgG molecule.

three independent protein moieties connected through a flexible linker (the hinge region), *Figure 6.1B*. Two of these moieties, referred to as Fab regions, are of identical structure and each expresses a specific antigen-binding site; the third, the Fc, expresses interaction sites for ligands that activate clearance mechanisms. These effector ligands include three structurally homologous cellular Fc receptor types (FcγRI, FcγRII, FcγRIII) (Ravetch, 2001; van Sorge, *et al.*, 2003; Woof and Burton, 2004), the C1q component of complement (Thommesen *et al.*, 2000) and the neonatal Fc receptor (FcRn) (Burmeister *et al.*, 1994; Vicarro *et al.*, 2005). Activation of Fc receptors and the C1q component of complement initiate inflammatory cascades that combat and resolve episodes of infection by pathogens. These activities are critically dependent on IgG-Fc glycosylation and, in part, on individual antibody glycoforms (Ferrara *et al.*, 2006; Niwa *et al.*, 2005; Umana *et al.*, 1999). In contrast, activation of FcRn, which determines antibody half-life and placental transport, is independent of glycosylation status. The constant regions of the four subclasses exhibit >95% sequence homology, although each expresses a unique profile of effector functions.

The IgG-Fc region is a homodimer comprised of covalent inter-heavy chain disulfide bonded hinge regions and noncovalently paired C_H3 domains; the C_H2 domains are glycosylated through covalent attachment of oligosaccharide at asparagine 297 (Asn-297). X-ray crystallographic analysis reveals discreet structure for the oligosaccharide that is integral to the IgG-Fc structure and forms multiple noncovalent interactions with the protein surface of the C_H2 domain; thus the protein and oligosaccharide exert reciprocal influences on the conformation of each other (Deisenhofer, 1981; Radeav *et al.*, 2001; Sondermann *et al.*, 2000). There is cumulative evidence that interaction sites on IgG-Fc for FcγRI, FcγRII, FcγRIII and C1q effector ligands are comprised of the protein moiety only. However, generation of the essential IgG-Fc protein conformation is dependent on the presence of the oligosaccharide. Thus, effector mechanisms mediated through FcγRI, FcγRII, FcγRIII and C1q are severely compromised or ablated for aglycosylated or deglycosylated forms of IgG (Krapp *et al.*, 2003; Mimura *et al.*, 2000, 2001; Pound *et al.*, 1993a,b).

6.7 IgG-Fc glycosylation

The oligosaccharide of normal polyclonal human IgG-Fc is of the diantennary complex type and shows considerable heterogeneity. A 'core' heptasaccharide can be defined with variable addition of outer arm sugar residues. A total of 32 different oligosaccharides may be attached generating, potentially, more than four hundred glycoforms, given random pairing of heavy chain glycoforms. Analysis of the oligosaccharide released from normal polyclonal IgG-Fc shows a paucity of sialylation (<10%) and 12 of the possible 16 neutral oligosaccharides to predominate, *Figure 6.3* (Jefferis *et al.*, 1990); providing the potential to generate a total of 72 glycoforms. These data are for polyclonal IgG that is the product of many thousands of individual plasma cell clones. Evidence from the glycosylation profile of monoclonal human IgG, produced by malignant plasma cell clones in multiple myeloma, suggest that although individual antibody populations are comprised of multiple glycoforms, each clone exhibits a unique

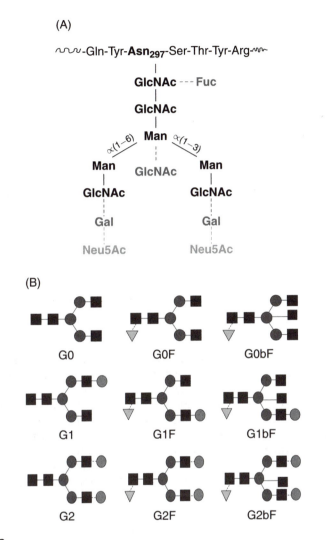

Figure 6.2

(A) Complex diantennary oligosaccharide: 'core' sugars are dark colored.
(B) Principle neutral oligosaccharides released from normal human serum IgG.
(■) N-acetylglucosamine (GalNAc); (●) mannose (Man); (●) galactose (Gal);
(▼) fucose (Fuc); Neu5Ac, N-acetylneuraminic acid.

oligosaccharide (glycoform) profile (Farooq *et al.*, 1997). Similarly, the glycoform profile of monoclonal mouse antibodies, produced by hybridomas, is heterogeneous but clone specific (Lund *et al.*, 1993). Determination of the glycoform profile of a rMAb product should be an early event in clone selection for a therapeutic rMAb.

Glycosylation is a co-translational PTM initiated by attachment of a glucosylated high mannose oligosaccharide (GlcNAc$_2$Man$_9$Glu$_3$), the glycosylated protein binds chaperones that aid folding fidelity and exercise quality

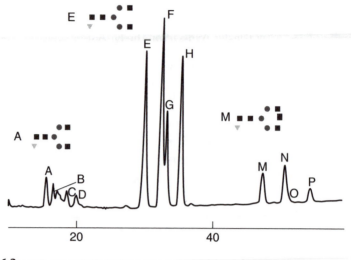

Figure 6.3

Neutral oligosaccharides released form polyclonal human IgG-Fc.

control. The oligosaccharide is trimmed to a $GlcNAc_2Man_9Glu$ structure (Petrescu *et al.*, 2004; Varki, 1999) that transits the Golgi apparatus where the oligosaccharide is initially trimmed, by glycosidases, to a $GlcNAc_2Man_5$ structure, prior to the action of successive glycosyltransferases, to generate complex diantennary structures (Ferrara *et al.*, 2006). Early studies established that CHO cells are able to produce rMAb IgG with the majority of glycoforms identical to those of polyclonal human IgG. However, under nonoptimal conditions CHO, NS0 and Sp2/0 cells can produce a number of abnormally glycosylated products that lack potency or are potentially immunogenic and unacceptable as therapeutics. Potential glycosylation sites may not be fully occupied, oligosaccharides may not be fully processed, resulting in the presence of high mannose glycoforms. In addition CHO and murine cells can also add sugars that are not found on normal human IgG and are known to be immunogenic, e.g. galactose α(1–3) galactose and *N*-glycolylneuraminic acid structures (Galili, 2005; Nguyen, 2005). A major achievement of the biopharmaceutical industry has been the development of cell lines that can be expanded in serum and protein free media to produce rMAb whilst maintaining glycosylation fidelity and minimizing the content of abnormal glycoforms. Regulatory authorities require that the glycoform profile of the marketed antibody product be maintained within strict limits, set by the glycoform profile of the material that gained licensing approval.

6.8 IgG-Fab glycosylation

It is established that 15–20% of polyclonal human IgG molecules bear N-linked oligosaccharides within the IgG-Fab region, in addition to the conserved glycosylation site at Asn 297 in the IgG-Fc (Dunn-Walters, 2000;

Holland *et al.*, 2006; Jefferis, 2005; Youings 1996;). There are no consensus sequences for N-linked oligosaccharide within the constant domains of either the kappa or lambda light chains or the C_H1 domain of heavy chains. Therefore, when present they are attached in the variable regions of the kappa (Vκ), lambda (Vλ) or heavy (V_H) chains; sometimes both. In the immunoglobulin sequence database ~20% of IgG V regions have N-linked glycosylation consensus sequences (Asn-X-Thr/Ser; where X can be any amino acid except proline). Interestingly, these consensus sequences are mostly not germline encoded but result from somatic mutation, suggesting positive selection for improved antigen binding (Jefferis, 2005; McCann *et al.*, 2006). The functional significance for IgG-Fab glycosylation of polyclonal IgG has not been fully evaluated but data emerging for monoclonal antibodies suggests that Vκ, Vλ or V_H glycosylation can have a neutral, positive or negative influence on antigen binding (Co et al., 1993; Jefferis, 2005;).

Analysis of polyclonal human IgG-Fab reveals the presence of diantennary oligosaccharides that are extensively galactosylated and substantially sialylated, in contrast to the oligosaccharides released from IgG-Fc (Holland *et al.*, 2006; Jefferis, 2005). This is somewhat surprising as one may expect that 'random' mutations would result in the generation of a glycosylation motifs at many sites within a polyclonal antibody population and that site-specific glycosylation would be observed with varied incorporation of high mannose, triantennary oligosaccharides etc. It might also be anticipated that IgG-Fab oligosaccharides would be exposed/accessible and, therefore, that glycoforms with terminal galactose residues would be catabolized through the liver, due to recognition by the asialoglycoprotein receptor. The stable composition of IgG-Fab glycoforms suggests that this is not the case. However, we cannot know whether more complex IgG-Fab glycoforms are generated but are rapidly removed and, therefore, not detected on analysis.

A similar glycoform profile was observed for human monoclonal IgG1 (myeloma) protein that is glycosylated within the V_L region. Whilst both the IgG-Fc and IgG-Fab bore diantennary structures the IgG-Fc oligosaccharides were predominantly nonfucosylated G0 and G1 whilst the IgG-Fab oligosaccharides were predominantly fucosylated, galactosylated, and sialylated (Jefferis, 2005). Interestingly, whilst the IgG-Fc oligosaccharides could be quantitatively released on exposure to PNGase F the IgG-Fab oligosaccharides were refractory, but could be released on exposure to Endo F, in contrast to the IgG-Fc oligosaccharides that could not be released with this enzyme (Jefferis, 2005). A similar pattern was observed for the rMAb therapeutic Cetuximab (Erbitux), which has specificity for the epidermal growth factor receptor and is licensed for the treatment of colon, head, and neck cancer etc. This rMAb, produced in Sp2/0 cells, bears an N-linked oligosaccharide at Asn 88 of the V_H region. Interestingly there is also a glycosylation motif at Asn 41 of the V_L but it is not occupied (Pendse *et al.*, 2004). Different glycoforms predominate at each site; the IgG-Fc bearing fucosylated G0 and G1 oligosaccharides and the IgG-Fab fucosylated G2 and sialylated oligosaccharides. Whilst the IgG-Fc oligosaccharides could be released on exposure to PNGase F the IgG-Fab oligosaccharide linkage was refractory. Thus, the rMAb and the Sp2/0 cell line appear to replicate the situation observed for polyclonal human IgG produced *in vivo*.

Other experiences of IgG-Fab glycosylation have been reported. A detailed analysis of the glycoforms of a humanized IgG rMAb, expressed in Sp2/0 cells, bearing oligosaccharides at Asn 56 of the V_H and Asn 297 reveals the expected IgG-Fc oligosaccharides profile of predominantly fucosylated G0 and G1. However, 11 oligosaccharides were released from the IgG-Fab, including triantennary, and other oligosaccharides not observed in normal human IgG (Huang *et al.*, 2005). Clearance rates, in mice, were independent of the IgG-Fc glycoform and of nine of the IgG-Fab oligosaccharides; marginal accelerated clearance of two oligosaccharides were observed. All IgG-Fab oligosaccharides were extensively sialylated with N-glycolylneuraminic acid, rather than N-acetylneuraminic acid. The IgG-Fab oligosaccharide linkages were refractory to release by PNGase F. Presentations made by biopharmaceutical companies at meetings have reported similar finding of conserved IgG-Fc oligosaccharide profiles and more heterogeneous IgG-Fab glycosylation (Pendse *et al.*, 2004); the latter showing relatively high levels of galactosylation and sialylation. It is important to emphasize that IgG-Fab glycosylation has not been shown to compromise clearance rates, at least in mouse models.

The influence of IgG-Fab glycosylation on antigen binding has been the subject of several reports. Three rMAbs with specificity for α(1–6)dextran, differing only in potential N-glycosylation sites at Asn 54, 58 or 60 in the V_H CDR2 region, were evaluated for antigen binding affinity. The Asn 54 and Asn 58 molecules, each bearing a complex diantennary oligosaccharide rich in sialic acid, were equivalent in antigen binding and the glycosylated forms had a 10- to 50-fold higher affinity for antigen compared with aglycosylated forms. In contrast, the Asn 60 molecule bore a high mannose oligosaccharide and had a lower affinity for antigen (Coloma *et al.*, 1999; Gala and Morrison, 2004). A significant proportion of IgG-Fab oligosaccharides bore Gal α(1–3) Gal structures (Endo et al., 1995). By contrast, humanization of a mouse anti-CD33 antibody with concomitant removal of a potential glycosylation site at Asn 73 of the V_H resulted in higher affinity for antigen; similarly deglycosylation of the original mouse antibody resulted in increased affinity (Co *et al.*, 1993). Increased affinity was reported for the deglycosylated form of a mouse antibody specific for ovomucoid bearing N-linked oligosaccharide in the light chain CDR2 (Fujimura *et al.*, 2000). A multi-specific human monoclonal antibody, produced in mouse-human heterohybridoma cells, has been reported to bear both diantennary and tetra-antennary oligosaccharides attached at Asn 75 of the V_H region and to include antigenic N-glycolylneuraminic acid sugar residues (Leibiger *et al.*, 1999).

The consistent observation of higher levels of galactosylation and sialylation for IgG-Fab N-linked oligosaccharides, in comparison to IgG-Fc, is thought to reflect increased exposure and/or accessibility. It might be expected, therefore, that the enzyme PNGase F would more readily cleave IgG-Fab oligosaccharides. However, to date, only contrary experiences have been reported. In conclusion, IgG-Fab glycosylation can impact differentially on the structural and functional characteristics of IgG. It may be exploited to increase the solubility and stability of antibodies, limiting aggregation and hence immunogenicity (Leibiger *et al.*, 1999). However, it offers a further challenge to the biopharmaceutical industry.

An intriguing feature of IgG-Fab glycosylation is being revealed in the study of human B-cell lymphoproliferative disease. Whilst ~10% of normal B cells bear surface Ig glycosylated within V_H or V_L the frequency amongst patients with sporadic Burkitt's lymphoma, endemic Burkitt's lymphoma and follicular lymphoma is 42%, 82%, and 94%, respectively (McCann, *et al.* 2006). Sequence analysis revealed multiple glycosylation motifs; one sequence encoding four V_H and two V_L motifs. These data are evidence of somatic mutation and an apparent selection for V_H/V_L glycosylation and could be related to the aetiology of the disease.

6.9 Cell engineering to influence glycoform profiles

Production CHO, NS0 and Sp2/0 cell lines yield rMAb having a restricted glycoform profile, relative to that observed for normal polyclonal human IgG, with G0F and G1F glycoforms predominating. Concern has been expressed for the relatively low levels of galactosylation and its possible impact on activation of the classical complement pathway; it has been reported that galactosylation impacts positively on the ability of Rituximab to lyse CD20 expressing cells (http://www.fda.gov/cder/biologics/review/ritugen112697-r2.pdf) but, to date, this appears to be a solitary experience.

Following an earlier paper suggesting that antibodies with oligosaccharides bearing a bisecting *N*-acetylglucosamine (peaks M, N, O, P; *Figure 6.3*) are more efficient at recruiting ADCC the company Glycart generated a CHO cell line transfected with the GNTIII enzyme and produced an anti-neuroblastoma antibody product bearing bisecting GlcNAc residues that exhibited a 15- to 20-fold improvement in ADCC (Ferrara *et al.*, 2006; Umana *et al.*, 1999). A similar improvement was reported by IDEC for the rMAb Rituximab that is licensed for the treatment of nonHodgkin's lymphoma and is being developed for other indications (Davies *et al.*, 2001). It has been demonstrated that the efficacy of Rituximab results from sensitization of CD20 expressing malignant B cells and the recruitment of FcγRIIIa expressing natural killer (NK) cells to induce ADCC mediated killing; the role of CDC is thought to less important for Rituximab, although it has been shown to be significant for other anti-CD20 antibodies (Teeling *et al.*, 2004). The density of the CD20 antigen on the B-cell surface is probably an important parameter and a paucity of expression may be linked with lack of clinical response. If the increased efficacy for *in vitro* killing of B cells observed for the Rituximab glycoform bearing bisecting *N*-acetylglucosamine translates to *in vivo* activity substantial gains may be anticipated and tumor cells with a relatively low level of CD20 expression may be killed, resulting in a higher clinical response rates; similarly clinical responses for tumors with a high level of CD20 expression may be achieved with reduced doses of the therapeutic.

An alternative glycoform of rMAb influencing ADCC effector function was reported from Genentech (see Chapter 8). A mutant CHO cell line (LEC 13) was employed that is deficient in the addition of fucose to the primary *N*-acetylglucosamine residue to produce nonfucosylated glycoforms of Herceptin. A 40- to 50-fold increase in the efficacy of FcγRIIIa mediated ADCC was reported and some improvement in binding to certain polymorphic forms of FcγRII but was without effect on binding to FcγRI or C1q

(Shields *et al.*, 2002). The LEC 13 cell line was not considered to be suitable for development as a production vehicle. A similar improvement in ADCC was reported for the nonfucosylated fraction of a recombinant anti-human IL-5 receptor (rhIL-5-R) antibody (Shinkawa *et al.*, 2003) produced in the rat-derived YB2/0 cell line. This cell line had previously been reported to express the GNTIII enzyme and to produce antibody expressing bisecting *N*-acetyl-glucosamine residues, which appeared to correlate with improved ADCC. Physical separation of the nonfucosylated and bisecting *N*-acetylglu-cosamine glycoforms suggested that it was the absence of fucose rather than the presence of bisecting *N*-acetylglucosamine that resulted in enhanced ADCC. The separation protocol has been applied to the isolation of nonfu-cosylated Rituximab produced in the YB2/0 cell line and improved ADCC has been correlated with increased affinity of the antibody for the FcγRIIIa receptor (Okazaki *et al.*, 2004). The company Biowa has established a fucosyl-transferase 'knock-out' cell line for routine production of nonfucosylated IgG antibodies (Ymane-Ohuki, *et al.*, 2004). Importantly they have estab-lished that FcγRIIIa mediated ADCC is amplified for the nonfucosylated glycoform of each of the IgG subclasses (Niwa *et al.*, 2005), in the whole molecule format and also as the truncated single chain Fv-Fc form (scFv-Fc)$_2$ (Natsume *et al.*, 2005). The increase in FcγRIIIa mediated ADCC has been correlated with an increase in the affinity of the antibody for the receptor and the nonfucosylated antibody overcomes the deficit evident for the FcγRIIIa Phe158 allotype.

6.10 IgG glycoforms and Fc effector functions

It is established that glycosylation of the IgG-Fc is essential for optimal expression of biological activities mediated through FcγRI, FcγRII, FcγRIII and the C1q component of complement (Jefferis, 2005; Woof and Burton, 2004). Present evidence suggests that it does not influence interactions with FcRn and consequently, presumably, the catabolic half-life or transport across the placenta; catabolic data for aglycosylated human IgG is only available from studies in a mouse model and the influence on placental transport in humans has not been reported. The IgG-Fc binding properties of bacterial proteins; for example, SpA, SpG are also unaffected (Deisenhofer, 1981; Jefferis, 2005; Jefferis *et al.*, 1998). The association constant of aglycosylated IgG1 or IgG3 binding to FcγRI is reduced by two orders of magnitude, relative to that observed for the normally glycosylated form. However, aglycosylated IgG3 antibody has been shown to mediate ADCC if a high level of target cell sensitization is achieved – that is, i.e. at high epitope density (Pound *et al.*, 1993a,b – and cellular activation through FcγRII and FcγRIII appears to be completely ablated (Jefferis *et al.*, 1998; Wright and Morrison, 1998). The association constant for C1q binding to aglycosylated IgG is also reduced by one order of magnitude and results in a complete loss of CDC (Lund *et al.*, 1996). Protein engineering, employing alanine scanning, has been used to 'map' amino acid residues deemed to be critical for FcγR and C1q binding. These studies 'map' the binding site for all four of these ligands to the hinge proximal or lower hinge region of the C$_H$2 domain (Idusogie, *et al.*, 2000; Lund *et al.*, 1996; Shields *et al.*, 2001).

X-ray crystallographic analysis of the IgG-Fc fragment, residues 216–446, reveals electron density for residues 238–443 only (Deisenhofer, 1981); thus the lower hinge region would appear to be mobile and without defined structure. This might appear to be incompatible with the suggestion that the lower hinge region is directly involved in the generation of structurally distinct interaction sites for the FcγR and C1q ligands. However, we have proposed that this region of the molecule is not without structure but is comprised of an equilibrium between multiple conformers, resulting from reciprocal interactions between the oligosaccharide and the protein moiety such that individual conformers are compatible with specific ligand recognition (Jefferis, 2005; Jefferis *et al.*, 1998). In the absence of the oligosaccharide a different set of conformers will be generated that are not compatible with ligand binding. X-ray crystallographic analysis of IgG-Fc in complex with a soluble recombinant form of the receptor, rFcγRIIIa, provided proof of the direct involvement of the lower hinge regions and hinge proximal C_H2 domain residues (Radaev *et al.*, 2001; Sondermann *et al.*, 2000).

The interaction site on the IgG-Fc is seen to include asymmetric binding to discrete conformations of the lower hinge residues of each heavy chain. A critical requirement is explained by this structure – that the IgG-Fc should be univalent for the FcγR. This is essential since if monomeric IgG were divalent it could cross-link cellular receptors and hence constantly activate inflammatory reactions. Interestingly, one structure reveals a possible contribution of the primary *N*-acetylglucosamine residue to binding whilst the other structure holds that there is no direct contact. Clearly, any contribution is minimal and the oligosaccharide contributes indirectly to the binding of these ligands. X-ray crystal structures of IgG-Fc in complex with SpA, FcRn, SpG and the autoantibody rheumatoid factor show that each of these ligands interacts with sites embracing residues of both the C_H2 and C_H3 domain, at their junction (Burmeister, 1994; Corper *et al.*, 1997; Deisenhofer, 1981; Sauer-Eriksson *et al.*, 1995) and the IgG-Fc is divalent for these ligands. It is evident, therefore, that the distribution of effector ligand binding and activation sites on IgG-Fc results from evolutionary selection for valency, amongst other properties.

There are two possible IgG-Fc functional activities for which direct binding to the oligosaccharide is essential – the mannan-binding lectin (MBL) and the cellular mannose receptor (MR) (Dong *et al.*, 1999; Malhotra *et al.*, 1995). Each of these lectins recognizes arrays of sugar residues that may be presented on the surface of microorganisms and they form a link between the innate and adaptive immune response. The sugar residues recognized include *N*-acetylglucosamine and there is evidence that immune complexes of G0F IgG can present arrays that bind and activate these lectins. Such activation may be presumed to be beneficial; however, there is evidence that immune complexes composed predominantly of G0 IgG can promote the adaptive immune response by uptake through the mannose receptor expressed on dendritic cells (Dong *et al.*, 1999). There is a possible downside to this activity for G0/G0F glycoforms of chimeric rMAb since they are potentially immunogenic and dendritic cells are the most efficient antigen presenting cell type. This route could similarly contribute to the observed development of human antibody responses to humanized or human (HAHA) rMAb therapeutics.

6.11 Glycosylation engineering

It will be evident that an ability to produce selected homogenous glycoforms of recombinant antibody molecules would be advantageous. Manipulation of culture conditions can have a limited, but significant, influence on the glycoform profile of product (Andersen *et al.*, 2000; Andersen and Reilly, 2000; Kunkel *et al.*, 2000; Yang and Butler, 2000) and may allow for manipulation of the glycoform profile over the time of a production run. Cell engineering is being undertaken in order to knock out and/or knock in genes encoding for selected glycosyltransferases; as illustrated above. Chapter 7 describes another example of this in the control of fucosylation of IgG. Subtle structural parameters also influence the glycosylation profile. Thus, alanine scanning studies showed that single amino acid replacements could result in gross changes in the glycoform profile of product; resulting in increased galactosylation and sialylation (Lund *et al.*, 1996). Similarly, radical differences in the glycoform profile were observed for a series of truncated IgG molecules (Lund *et al.*, 2000). It is also possible to modify the glycoform profile, *in vitro*, using glycosidases or glycosyltransferases and activated sugar precursors. We used this approach to determine the minimal oligosaccharide structure that could provide both structural stability and biological function for IgG-Fc (Krapp *et al.*, 2003; Mimura *et al.*, 2000, 2001b). The study showed that the initial GlcNAc-GlcNAc-Man trisaccharide conferred significant stability and activity in comparison with the aglycosylated form.

Changes in the glycoform profiles of polyclonal IgG have been reported that are characteristic for a number of inflammatory autoimmune diseases (Axford *et al.*, 2003; Holland *et al.*, 2006). These studies have focused on IgG-Fc glycosylation but studies comparing oligosaccharide profiles of IgG-Fab and IgG-Fc fragments reveal contrasting oligosaccharide profiles (Holland *et al.*, 2006; Youings *et al.*, 1996). Gross hypogalactosylation of IgG isolated from the serum of patients with systemic vasculitis experiencing an acute inflammatory episode was reported (Holland *et al.*, 2002). Subsequently, the oligosaccharide profiles of IgG-Fc, IgG-Fab and IgG-F(ab′)$_2$ and generated from the serum IgG revealed hypogalactosylation to be restricted to IgG-Fc, whilst IgG-Fab oligosaccharides are both galactosylated and sialylated (Holland *et al.*, 2006). This provides evidence that the glycosylation machinery is intact and demonstrates site specificity for the elongation of the oligosaccharide chains of diantennary complex structures. A significant feature of this finding is that it applies to the total IgG population. Two possible mechanisms may be contemplated: (i) following secretion IgG-Fc galactose residues are selectively removed by a specific galactosidase; (ii) a change in the micro-environment within the Golgi apparatus results in a conformational change within the IgG-Fc such that it is relatively inaccessible to the β-(1–4) galactosyltransferase enzyme. If the latter should prove to be the case there may be potential to establish similar conditions for recombinant antibody production cell lines offering consequent control of glycoform product.

6.12 Pharmacokinetics and placental transport

The catabolic half-life of human IgG is very extended, relative to other serum proteins; IgG1, IgG2 and IgG4 proteins having a half-life of ~21 days and IgG3 proteins ~7 days. Catabolism is mediated through the unique neonatal Fc receptor, FcRn, pathway. The receptor is expressed on the membrane of tissues widely expressed throughout the body. Cells continuously sample their environment, in part, through a process of membrane invagination with the formation of vacuoles, followed by fusion with membrane and re-expression. Fluid within the vacuole is acidified to pH ~6.5 and at this pH IgG present in the fluid phase binds and saturates FcRn. Excess IgG remaining in the fluid phase is susceptible to enzymatic cleavage whilst that bound to FcRn is protected. When the vacuole fuses with the membrane the FcRn-IgG complex is exposed to serum/fluid and is released at physiological pH, 7.2.

An X-ray crystal structure of IgG-Fc in complex with recombinant FcRn has shown the binding site to be at the C_H2/C_H3 interphase with residues of both domains contributing to binding (Burmeister *et al.*, 1994). Histidine residues within the site are crucial as their pK corresponds to the binding/release at pH 6.5 and 7.2, respectively. Interestingly, histidine 435 present in IgG1, IgG2, and IgG4 is replaced by arginine in the IgG3 of most population groups and could be related to the shorter half-life for this IgG subclass. An allotypic variant of IgG3, present within mongoloid populations, has histidine at 435. However, definitive data on its catabolic half-life is not available. The binding site on IgG-Fc for SpA overlaps that of FcRn and whilst IgG3-Arg435 does not bind SpA the IgG3-His435 variant does. Extensive mutational studies have identified residues crucial to FcRn binding and mutants with higher binding affinities that exhibit extended half-lives (Vicarro *et al.*, 2005; Zhou *et al.*, 2003). These studies may provide a route to optimizing the *in vivo* half-life of an antibody for a given therapeutic application. Importantly, interactions with FcRn are independent of the presence or absence of the oligosaccharide moiety at Asn 297.

The long half-life of IgG antibodies is being exploited through the generation of fusion proteins, e.g. single chain Fv-Fc (scFv-Fc)$_2$ (Natsume *et al.*, 2005), cytokine-IgG-Fc (http://www.syntnx.com/home.php). The presence of the IgG-Fc region contributes to improved stability, pharmacokinetics, and pharmacodynamics. A further development opens a new route for administration. It has been shown that FcRn is expressed in the central and upper airways and that drug–IgG-Fc fusion proteins delivered to these sites can be transferred, by transcytosis, to the systemic circulation. This is an exciting development with considerable promise and significance (http://www.syntnx.com/home.php; Dumont *et al.*, 2005).

Placental transport is effected through a complex series of events, as proteins have to undergo transcytosis through numerous cell types. Transfer across the syncytiotrophoblast of the chorionic villi is FcRn-mediated although other FcγR may be involved in further cellular transport (Simister, 2003). The influence of glycoform on this overall transportation process is not known.

6.13 Antibody therapeutics of the IgA class

It is our common experience that we are most prone to infection at mucosal surfaces, and the respiratory, gastrointestinal, and reproductive tracts. These sites present live tissue bathed in fluid (mucus) at body temperature, the ideal environment for colonization by commensal and pathogenic organisms. These mucosal surfaces are protected by the local mucosal immune system that is comprised of IgA antibodies that are subject to transcytosis through epithelial cells (Woof and Mestecky 2005); whilst IgG equilibrates with extravascular fluids it cannot cross the epithelial barrier to exert immune protection at these external sites. For indications requiring topical or oral administration, therefore rMAb of the IgA class would be the logical choice. The IgA present at mucosal surfaces is referred to as secretory IgA (sIgA) and is comprised of dimeric IgA in complex with a J chain and secretory component. The four-chain IgA molecule and J chain are produced in plasma cells, underlying the mucosal surface, and dimerize through disulfide bond formation with the J chain. The secreted IgA dimer binds to the polyIg receptor expressed on the apical surface of epithelial cells, is internalized, undergoes transcytosis to be expressed on the basal surface of the epithelial cells where a membrane-bound proteolytic enzyme cleaves the receptor to release the IgA dimer in complex with a residue of the polyIg receptor (secretory component). The production of recombinant secretory IgA is thus a challenge.

In humans there are two subclasses of IgA (IgA1 and IgA2) that differ structurally and functionally. The IgA1 subclass has an extended hinge region that is vulnerable to proteolytic enzymes produced by certain pathogenic bacteria; in contrast, the IgA2 subclass has a truncated hinge region that is not susceptible to these enzymes. The IgA1/IgA2 ratio varies throughout the gastrointestinal tract with sIgA1 predominating in the oral cavity and sIgA2 in the hostile environment of the lower gut. The sIgA2 subclass would seem to be the natural choice for the development of recombinant secretory IgA antibodies.

The secretory form of human IgA2 is complex in both structure and function. There are two allotypes of IgA2 that differ in sequence and the number of N-linked oligosaccharide moieties; both the J chain and secretory component are also glycosylated. There is evidence that some of the oligosaccharide structures are lectins for structures expressed by bacteria and that the binding of sIgA to these structures inhibits them binding to cell surface structures and hence colonisation. In spite of these apparent complications sIgA has been successfully produced in tobacco plants and one therapeutic is at an advanced stage of clinical evaluation (Wycoff, 2005). Whilst plant produced glycoproteins will bear nonhuman oligosaccharides it is anticipated that, since individuals are exposed to them in their diet, they will not cause problems when administered topically or orally; systemic administration of plant-produced biopharmaceuticals is more problematic and it is anticipated that humanization of the glycosylation machinery will be essential (Brooks, 2004; Gomord et al., 2005).

6.14 Non-antibody recombinant (glyco)protein therapeutics, 'biosimilar', and 'follow-on' biologics

This section is not a comprehensive review but is intended to illustrate issues likely to be encountered in the development of 'biosimilar' or 'follow-on' biologics and the concept of generics in this field (Combe et al., 2005).

Proteins and vaccines are classed as 'biologics' and the FDA has different rules for them as apposed to small molecule 'drugs'. Thus, any change in the production process for a biologic is considered to have a possible impact on product potency. Therefore, the concept of 'comparability' has been introduced as a pragmatic evaluation:

The FDA site (http://www.fda.gov/cder/guidance/compare.htm) states in terms of comparability testing, manufacturers should generally perform extensive analytical testing complemented by functional testing if manufacturing changes occur in the process of producing the bulk drug substance. Examples of such changes include the following: a change in manufacturing site; modifications to cell or seed strains, including changes to the master cell bank; fermentation; and isolation or purification. In some cases, complementary pharmacology data or biologic response data (e.g., antibody titers for vaccines) may be needed.

It concludes:

FDA may determine that manufacturers of biological products, including therapeutic biotechnology-derived products regulated as biologics or drugs, may make manufacturing changes without conducting additional clinical efficacy studies if comparability test data demonstrate to FDA that the product after the manufacturing change is safe, pure, potent and effective.

The debate over 'follow-on' biologics centers on the complexity of a protein/glycoprotein and the manufacturing process being such that the product of one company can never be exactly reproduced in another facility. If sustained there would be a requirement for 'follow-on' biologics to be subject to new clinical trials before they could be licensed, which would compromise the potential cost advantage.

6.14.1 Erythropoietin

Recombinant erythropoietin (rEPO) is a 'blockbuster' drug with annual sales exceeding $10 billion. The patent for the original product expired in 2004 and it is a prime target for exploration of the legal and regulatory issues relating to 'follow-on'/generic products. Natural EPO, derived from urine, is a glycoprotein bearing three N-linked and one O-linked oligosaccharides, which account for a significant proportion of its mass (~40%) and contribute to biologic efficacy. In early studies, rEPO, produced in CHO cells, was shown to have a somewhat higher functional activity *in vitro* compared to endogenous EPO. *In vivo*, however, rEPO was essentially without therapeutic benefit. It was established that rEPO lacked terminal sialic acid residues and that the exposed terminal galactose residues resulted in accelerated clearance by the liver, mediated through the asialylglycoprotein receptor. Consequently, the rEPO was fractionated and only the sialylated glycoprotein employed as the licensed therapeutic. Whilst this was a somewhat expensive and wasteful procedure but could be sustained because of the low therapeutic dose required. Several variant forms of rEPO have been developed, the most recent being Aranesp® (darbepoetin alfa) which has been shown to have improved pharmacokinetics and *in vivo* biologic activity. Two additional N-linked glycosylation motifs were engineered into the EPO gene sequence to produce Aranesp having five N-linked oligosaccharide moieties and consequent increased sialylation (Egrie *et al.*, 2003).

6.14.2 Tissue-type plasminogen activator

Naturally produced human tissue-type plasminogen activator (t-PA) expresses one N-linked glycosylation motif that exhibits variable occupancy. The activity of aglycosylated t-PA compromises specific activity and fibrin binding. A similar variation in site occupancy is reported for recombinant t-PA produced in CHO cells. Interestingly, it was shown that site occupancy varied according to the growth rate of the cells, with slow growth rates resulting in higher levels of occupancy (Andersen *et al.*, 2000). This suggests that following division a finite time is required to establish the glycosylation machinery and cell selection for division time could be a relevant parameter.

6.14.3 Granulocyte-macrophage colony stimulating factor (GM-CSF)

The functional activity and serum half-life of this cytokine is dependent on appropriate glycosylation at two N-linked and one O-linked oligosaccharides (Forno *et al.*, 2004). The recombinant product, produced in CHO cells, has been shown to be heterogeneous with variable site occupancy and consequently the level of sialylation. The aglycosylated and asialylated proteins have short half-lives and may account for the observed immunogenicity. Differences in glycoform profiles have been reported for products of the same CHO cell line depending on culture conditions and whether they were expanded in suspension or adherent cell cultures (Forno *et al.*, 2004).

6.14.4 Granulocyte-colony stimulating factor

The natural form of this cytokine bears a single O-linked oligosaccharide; recombinant granulocyte-colony stimulating factor (G-CSF) produced in *E. coli* and CHO cells are available as licensed products and appear to be equivalent in *in vitro* assays. It has been reported, however, that the nonglycosylated form, produced in *E. coli*, is inactivated when incubated in human serum. It was subsequently demonstrated that inactivation is due to proteolysis and that the glycosylated CHO cell product is resistant to the proteolytic process (Carter *et al.*, 2004).

6.14.5 Activated protein C

Protein C plays a central role in the regulation of vascular function. It is produced as a zymogen that is converted to activated protein C (APC) following proteolytic cleavage of a 12-amino acid peptide. In addition to four N-linked glycosylation sites the activity of recombinant human APC (rhAPC) is dependent on the first nine glutamic acid residues being converted to γ-carboxyglutamates and one asparagine residue being converted to ε-β-hydroxyaspartate. The site of APC biosynthesis is the liver, although two liver cell lines (HepG2 & FAZA) failed to produce fully active rhAPC; CHO, BHK and other cell lines similarly failed to produce active rhAPC. Therapeutically active rhAPC was produced by a HEK293 cell line and required the development of new expression vectors and methods for the efficient secretion of rhAPC from HEK293 cells (Grinnell *et al.*, 2005).

6.15 Conclusions

It will be abundantly clear from the foregoing that each candidate protein therapeutic is unique and must be investigated on a case-by-case basis. It is evident that a universal cell line for the production of recombinant protein therapeutics is not available and that engineering a cell line to optimize one product (e.g. antibodies) may reduce applicability to the production of other products. This review has focused on glycosylation but opened with the reminder that >350 PTMs have been characterized and that individual molecules may bear the stamp of multiple PTM events. In addition whilst we focus on the known biologic activity of a given protein therapeutic each may exercise several roles *in vivo* (i.e. are pleomorphic) and each may be influenced by particular PTMs. The present and future challenge is to develop means of monitoring mechanisms activated *in vivo* and individual variations determined by the interaction of multiple polymorphisms – I believe it is called 'systems biology'.

References

Alete DE, Racher AJ, Birch JR, Stansfield SH, James DC and Smales CM (2005) Proteomic analysis of enriched microsomal fractions from GS-NS0 murine myeloma cells with varying secreted recombinant monoclonal antibody productivities. *Proteomics* **18**: 4689–4704.

Andersen DC and Reilly DE (2000) Production technologies for monoclonal antibodies and their fragments. *Curr Opin Biotechnol* **15**: 456–462.

Andersen D, Bridges T, Gawlitzek M and Hoy C (2000) Multiple cell culture factors can effect the glycosylation of ASN-184 in CHO-produced tissue-type plasminogen activator. *Biotechnol Bioeng* **70**: 25–31.

Axford JS, Cunnane G, Fitzgerald O, Bresnihan B and Frears ER (2003) Rheumatic disease differentiation using immunoglobulin G sugar printing by high-density electrophoresis. *J Rheumatol* **12**: 2540–2546.

Bauman M and Meri S (2004) Techniques for studying protein heterogeneity and post-translational modifications. *Expert Rev Proteomics* **2**: 207–217.

Birch JR and Racher AJ. (2006) Antibody Production. Advanced Drug Delivery Reviews, 58, 671–685.

Birch J (2005) Upstream mammalian cell processing – challenges and prospects. Bioprocess International presentation, Berlin.

Brooks SA (2004) Appropriate glycosylation of recombinant proteins for human use: implications of choice of expression system. *Mol Biotechnol* **28**: 241–255.

Burmeister WP, Huber AH and Bjorkman PJ (1994) Crystal structure of the complex of rat neonatal Fc receptor with Fc. *Nature* **372**: 379–383.

Butler M, Quelhas D, Critchley AJ *et al.* (2003) Detailed glycan analysis of serum glycoproteins of patients with congenital disorders of glycosylation indicates the specific defective glycan processing step and provides an insight into pathogenesis. *Glycobiology* **13**: 601–622.

Co MS, Scheinberg DA, Avdalovic NM, McGraw K, Vasquez M, Caron PC and Queen C (1993) Genetically engineered deglycosylation of the variable domain increases the affinity of an anti-CD33 monoclonal antibody. *Mol Immunol* **30**: 1361–1367.

Coloma MJ, Trinh RK, Martinez AR and Morrison SL (1999) Position effects of variable region carbohydrate on the affinity and in vivo behavior of an anti-(1→6) dextran antibody. *J Immunol* **162**: 2162–2170.

Combe C, Tredree RL and Schellekens H (2005) Biosimilar epoetins: an analysis based on recently implemented European medicines evaluation agency guidelines on comparability of biopharmaceutical proteins. *Pharmacotherapy* **25**: 954–962.

Corper AL, Sohi MK, Bonagura VR, Steinitz M, Jefferis R, Feinstein A, Beale D, Taussig MJ and Sutton BJ (1997) Structure of human IgM rheumatoid factor Fab bound to its autoantigen IgG Fc reveals a novel topology of antibody–antigen interaction. *Nat Struct Biol* **4**: 374–381.

Davies J, Jiang L, LaBarre MJ, Anderson D and Reff M (2001) Expression of GTIII in a recombinant anti-CD20 CHO production cell line: Expression of antibodies of altered glycoforms leads to an increase in ADCC thro' higher affinity for FcRIII. *Biotech Bioeng* **74**: 288–294.

Deisenhofer J (1981) Crystallographic refinement and atomic models of a human Fc fragment and its complex with fragment B of protein A from *Staphylococcus aureus* at 2.9- and 2.8-Å resolution. *Biochemistry* **20**: 2361–2370.

Dong X, Storkus WJ and Salter RD (1999) Binding and uptake of agalactosyl IgG by mannose receptor on macrophages and dendritic cells. *J Immunol* **163**: 5427–5434.

Dumont JA, Bitonti AJ, Clark D, Evans S, Pickford M and Newman SP (2005) Delivery of an erythropoietin-Fc fusion protein by inhalation in humans through an immunoglobulin transport pathway. *J Aerosol Med* **18**: 294–303.

Dunn-Walters DK, Boursier L and Spencer J (2000) Effect of somatic hypermutation on potential N-glycosylation sites in human immunoglobulin heavy chain variable regions. *Mol Immunol* **37**: 107–113.

Egrie JC, Dwyer E, Browne JK, Hitz A and Lykos MA (2003) Darbepoetin alfa has a longer circulating half-life and greater in vivo potency than recombinant human erythropoietin. *Exp Hematol* **31**: 290–299.

Endo T, Wright A, Morrison SL and Kobata A. Glycosylation of the variable region of immunoglobulin G—site specific maturation of the sugar chains. Mol Immunol. 1995 **32**: 931–40.

Farooq M, Takahashi N, Arrol H, Drayson M and Jefferis R (1997) Glycosylation of antibody molecules in multiple myeloma. *Glycoconjugate J* **14**: 489–492.

Ferrara C, Brunker P, Suter T, Moser S, Puntener U and Umana P (2006) Modulation of therapeutic antibody effector functions by glycosylation engineering: Influence of Golgi enzyme localization domain and co-expression of heterologous beta1, 4-N-acetylglucosaminyltransferase III and Golgi alpha-mannosidase II. *Biotechnol Bioeng* [E-pub ahead of print, 24 January 2006].

Forno G, Bollati Fogolin M, Oggero M, Kratje R, Etcheverrigaray M, Condrat HS and Nimtz M (2004) Recombinant human GM-CSF. 127 AA (14 kDa) monomeric protein with two glycosylation sites. *Eur J Biochem* **271**: 907–919.

Freeze HH (2002) Human disorders in N-glycosylation and animal models. *Biochim Biophys Acta* **1573**: 388–393.

Fujimura Y, Tachibana H, Eto N and Yamada K (2000) Antigen binding of an ovomucoid-specific antibody is affected by a carbohydrate chain located on the light chain variable region. *Biosci Biotechnol Biochem* **64**: 2298–2305.

Gala FA and Morrison SL (2004) V region carbohydrate and antibody expression. *J Immunol* 172: 5489–5494.

Galili U (2005) The alpha-gal epitope and the anti-Gal antibody in xenotransplantation and in cancer immunotherapy. *Immunol Cell Biol* 83: 674–686.

Gomord, V., Chamberlain, P., Jefferis, R. and Foye, L. (2005) Biopharmaceutical production in plants: problems, solutions and opportunities. Trends Biotechnol. 23: 559–65.

Grinnell BW, Yan SB and Macias WL (2005) ACTIVATED PROTEIN C. In: *Directory of Therapeutic Enzymes* (eds B.M. McGrath and G. Walsh). Taylor & Francis, Boca Raton, FL, pp. 69–95.

Gu J, Kondo A, Okamoto N, Wada Y (1994) Oligosaccharide structures of immunoglobulin G from two patients with carbohydrate-deficient glycoprotein syndrome. *Glycosylation Disease* 1: 247–252.

Hadley AG, Zupanska B, Kumpel BM, Pilkington C, Griffiths HL, Leader KA, Jones J, Booker DJ and Sokol RJ (1995) The glycosylation of red cell autoantibodies affects their functional activity in vitro. *Br J Haematol* 91: 587–594.

Harris RJ (2005) Heterogeneity of recombinant antibodies: linking structure to function. *Dev Biol* 122: 117–127.

Holland M, Takada K, Okumoto T *et al.* (2002) Hypogalactosylation of serum IgG in patients with ANCA-associated systemic vasculitis. *Clin Exp Immunol* 129: 183–190.

Holland M, Yagi H, Takahashi N, Kato K, Goodall DM, Savage COS and Jefferis R (2006) Differential glycosylation of polyclonal IgG, IgG-Fc and IgG-Fab isolated from the sera of patients with ANCA associated systemic vasculitis. *BBA – General Subjects* [E-pub ahead of print, 27 December 2005].

Huang L, Biolosi S, Bales KR and Kuchibhotla U Impact of variable domain glycosylation on antibody clearance: An LC/MS characterization. Anal Biochem. 2006 349: 197–207.

http://www.syntnx.com/home.php

http://us.expasy.org/sprot/hpi/

http://www.researchd.com/rdikits/rdisubbk.htm

http://www.fda.gov/cder/biologics/review/ritugen112697-r2.pdf

Idusogie EE, Presta LG, Gazzano-Santoro H, Totpal K, Wong PY, Ultsch M, Meng YG and Mulkerrin MG (2000) Mapping of the C1q binding site on rituxan, a chimeric antibody with a human IgG1 Fc. *J Immunol* 164: 4178–4184.

Ito K, Takahashi N and Hirayama M (1993) Abnormalities in the oligosaccharide moieties of immunolglobulin G in patients with mytonic dystrophy. *J Clin Biochem Nutr* 14: 61–69.

Jefferis R (2005) Glycosylation of recombinant antibody therapeutics. *Biotechnol Prog* 21: 11–16.

Jefferis R, Lund J, Mizutani H, Nakagawa H, Kawazoe Y, Arata Y and Takahashi N (1990) A comparative study of the N-linked oligosaccharide structures of human IgG subclass proteins. Biochem J. 268: 529–37.

Jefferis R, Lund J and Pound JD (1998) IgG-Fc mediated effector functions: molecular definition of interaction sites for effector ligands and the role of glycosylation. *Immunol Rev* 163: 59–76.

Kohler G and Milstein C (1975) Continuous cultures of fused cells secreting antibody of predefined specificity. *Nature* 256: 495–497.

Krapp S, Mimura Y, Jefferis R, Huber R and Sondermann P (2003) Structural analysis of human IgG glycoforms reveals a correlation between oligosaccharide content, structural integrity and Fcγ-receptor affinity. *J Mol Biol* 325: 979–989.

Kumpel BM Monoclonal anti-D development programme. (2002) Transpl Immunol. 10: 199–204.

Kunkel JP, Jan DC, Butler M and Jamieson JC (2000) Comparisons of the glycosylation of a monoclonal antibody produced under nominally identical cell culture conditions in two different bioreactors. *Biotechnol Prog* **16**: 462–470.

Lapolla A, Fedele D and Traldi P (2001) Diabetes and mass spectrometry. *Diabetes Metab Res Rev* **17**: 99–112.

Leibiger H, Wustner D, Stigler RD and Marx U (1999) Variable domain-linked oligosaccharides of a human monoclonal IgG: structure and influence on antigen binding. *Biochem J* **338**: 529–538.

Lund J, Takahashi N, Nakagawa H, Bentley T, Hindley S, Tyler R, Goodall M and Jefferis R (1993) Control of IgG/Fc glycosylation: a comparison of oligosaccharides from chimeric human/mouse and mouse subclass immunoglobulin Gs. *Mol Immunol* **30**: 741–748.

Lund J, Takahashi N, Pound J, Goodall M and Jefferis R (1996) Multiple interactions of IgG with its core oligosaccharide can modulate recognition by complement and human FcγRI and influence the synthesis of its oligosaccharide chains. *J Immunol* **157**: 4963–4969.

Lund J, Takahashi N, Popplewell A, Goodall M, Pound J, Tyler R, King D and Jefferis R (2000) Expression and characterisation of truncated glycoforms of humanised L243 IgG1: architectural features can influence synthesis of its oligosaccharide chains and affect superoxide production triggered through human FcγRI. *Eur J Biochem* **267**: 7246–7257.

Malhotra R, Wormald MR, Rudd PM, Fischer PB, Dwek RA and Sim RB (1995) Glycosylation changes of IgG associated with rheumatoid arthritis can activate complement via the mannose-binding protein. *Nat Med* **1**: 237–243.

Mann M and Jensen ON (2003) Proteomic analysis of post-translational modifications. *Nat Biotechnol* **21**: 255–261.

McCann KJ, Johnson PW, Stevenson FK and Ottensmeier CH (2006) Universal N-glycosylation sites introduced into the B-cell receptor of follicular lymphoma by somatic mutation: a second tumorigenic event? *Leukemia* **20**: 530–534.

Mimura Y, Church S, Ghirlando R, Dong S, Goodall M, Lund J and Jefferis R (2000) The influence of glycosylation on the thermal stability and effector function expression of human IgG1-Fc: properties of a series of truncated glycoforms. *Mol Immunol* **37**: 697–706.

Mimura Y, Lund J, Church S, Dong S, Li J, Goodall M and Jefferis R (2001a) Butyrate increases production of human chimeric IgG in CHO-K1 cells whilst maintaining function and glycoform profile. *J Immunol Methods* **247**: 205–216.

Mimura Y, Sondermann P, Ghirlando R, Lund J, Young SP, Goodall M and Jefferis R (2001b). The role of oligosaccharide residues of IgG1-Fc in FcγIIb binding. *J Biol Chem* **276**: 45539–45547.

Mirik GR, Bradt BM, Denardo SJ and Denardo GL (2004) A review of human anti-globulin antibody (HAGA, HAMA, HACA, HAHA) responses to monoclonal antibodies. Not four letter words. *Q J Nucl Med Mol Imaging* **48**: 251–257.

Natsume A, Wakitani M, Yamane-Ohnuki N, Shoji-Hosaka E, Niwa R, Uchida K, Satoh M and Shitar K (2005) Fucose removal from complex-type oligosaccharide enhances the antibody-dependent cellular cytotoxicity of single-gene-encoded antibody comprising a single-chain antibody linked the antibody constant region. *J Immunol Methods* **306**: 93–103. [E-pub 3 October 2005].

Nezlin R and Ghetie V (2004) Interactions of immunoglobulins outside the antigen-combining site. *Adv Immunol* **82**: 155–215.

Nguyen DH, Tanqveranuntakul P and Varki A. Effects of natural human antibodies against a nonhuman sialic acid that metabolically incorporates into activated and malignant immune cells. J Immunol. 2005 **175**: 228–36.

Niwa R, Natsume A, Uehara A, Wakitani M, Iida S, Uchidqa K, Satoh M and Shitara K (2005) IgG subclass-independent improvement of antibody-dependent cellular cytotoxicity by fucose removal from Asn297-linked oligosaccharides. *J Immunol Methods* **306**: 151–160. [E-pub 22 September 2005].

O'Donovan C, Apweiler R and Bairoch A (2001) The human proteomics initiative (HPI). *Trends Biotechnol* **19**: 178–181.

Okazaki A, Shoji-Hosaka E, Nakamura K, Wakitani M, Uchida K, Kakita S, Tsumoto K, Kumagai I and Shitara K (2004) Fucose depletion from human IgG1 oligosaccharide enhances binding enthalpy and association rate between IgG1 and FcγRIIIa. *J Mol Biol* **336**: 1239–1249.

Parekh RB, Dwek RA, Sutton BJ *et al.* (1985) Association of rheumatoid arthritis and primary osteoarthritis with changes in the glycosylation pattern of total serum IgG. *Nature* **316**: 452–457.

Pendse GJ, Katarzyna C, DelMauro L *et al.* (2004) Process development and scale-up of Cetuximab production process. ImClone Systems Incorporated, USA Cell Culture Engineering IX, Cancun, Mexico.

Petrescu AJ, Milac AL, Petrescu SM, Dwek RA and Wormald MR (2004) Statistical analysis of the protein environment of N-glycosylation sites: implications for occupancy, structure, and folding. *Glycobiology* **14**: 103–114. [E-pub 26 September 2003].

Poland DC, Vallejo JJ, Niedden HW, Nijmeyer R, Calafat J, Hack CE, Van Het Hof B and van Dijk W (2005) Activated human PMN synthesize and release a strongly fucosylated glycoform of {alpha}1-acid glycoprotein, which is transiently deposited in human myocardial infarction. *J Leukoc Biol* **78**: 453–461.

Pound JD, Lund J, Jones PT, Winter G and Jefferis R (1993a) FcγRI mediated triggering of the human mononuclear phagocyte respiratory burst. *Mol Immunol* **30**: 233–241.

Pound JD, Lund J and Jefferis R (1993b) Aglycosylated chimeric human IgG3 can trigger the human phagocyte respiratory burst. *Mol Immunol* **30**: 469–478.

Radaev S, Motyka S, Fridman WH, Sautes-Fridman C and Sun PD (2001) The structure of human type III Fcγ receptor in complex with Fc. *J Biol Chem* **276**: 16469–16477.

Raju TS, Briggs JB, Borge SM and Jones AJ (2000) Species-specific variation in glycosylation of IgG: evidence for the species-specific sialylation and branch-specific galactosylation and importance for engineering recombinant glycoprotein therapeutics. *Glycobiology* **10**: 477–486.

Ravetch JV and Bolland S (2001) IgG Fc receptors. *Annu Rev Immunol* **19**: 275–290.

Rodriguez J, Spearman M, Huze L and Butler M (2005) Enhanced production of monomeric interferon-beta by CHO cells through the control of culture conditions. *Biotechnol Prog* **21**: 22–30.

Sauer-Eriksson AE, Kleywegt GJ, Uhl M and Jones TA (1995) Crystal structure of the C2 fragment of streptococcal protein G in complex with the Fc domain of human IgG. *Structure* **3**: 265–278.

Shields RL, Lai J, Keck R, O'Connell LY, Hong K, Meng YG, Weikert SH and Presta LG (2001) Lack of fucose on human IgG1 N-linked oligosaccharide improves binding to human Fcgamma RIII and antibody-dependent cellular toxicity. *J Biol Chem* **277**: 26733–26740.

Shinkawa T, Nakamura K, Yamane N *et al.* (2003) The absence of fucose but not the presence of galactose or bisecting *N*-acetylglucosamine of human IgG1 complex-type oligosaccharides shows the critical role of enhancing antibody-dependent cellular cytotoxicity. *J Biol Chem* **278**: 3466–3473.

Simister NE (2003) Placental transport of immunoglobulin G. *Vaccine* **21**: 3365–3369.

Sinclair, AM and Elliott, S (2005) Glycoengineering: The effect of glycosylation on the properties of therapeutic proteins. *J Pharm Sci* **94**: 1626–1635.

Smalling R, Foot M, Molineux G, Swanson SJ and Elliott S (2004) Drug-induced and antibody-mediated pure red cell aplasia: A review of literature and current knowledge. *Biotechnol Annu Rev* 10: 237–250.

Sondermann P, Huber R, Oosthuizen V and Jacob U (2000) The 3.2-A crystal structure of the human IgG1 Fc fragment-FcγRIII complex. *Nature* 406: 267–273.

Takahashi N, Ishii I, Ishihara H, Mori M, Tejima S, Jefferis R, Endo S and Arata Y (1987) Comparative structural study of the N-linked oligosaccharides of human normal and pathological immunoglobulin G. *Biochemistry* 26: 1137–1144.

Tanner JF (2005) Designing antibodies for oncology. *Cancer Metastasis Rev* 24: 585–598.

Teeling JL, French RR, Cragg MS *et al.* (2004) Characterization of new human CD20 monoclonal antibodies with potent cytolytic activity against non-Hodgkin lymphomas. *Blood* 104: 1793–1800.

Thommesen JE, Michaelsen TE, Loset GA, Sandie I and Brekke OH (2000) Lysine 322 in the human IgG3 CH2 domain is crucial for antibody dependent complement activation. *Mol Immunol* 37: 995–1004.

Umana P, Jean-Mairet J, Moudry R, Amstutz H and Bailey J E (1999) Engineered glycoforms of an antineuroblastoma IgG1 with optimized antibody-dependent cellular cytotoxic activity. *Nat Biotechnol* 17: 176–180.

Van den Nieuwenhof IM, Koistinen H, Easton RL *et al.* (2000) Recombinant glycodelin carrying the same type of glycan structures as contraceptive glycodelin-A can be produced in human kidney 293 cells but not in chinese hamster ovary cells. *Eur J Biochem* 267: 4753–4762.

Van Sorge NM, van der Pol WL and van de Winkel JG. FcgammaR polymorphisms: Implications for function, disease susceptibility and immunotherapy. Tissue Antigens. 2003 61: 189–202.

Varki A (1999) This is a text book to which Varki has contributed several chapters in: *Essentials of Glycobiology* (eds A. Varki *et al.*). Cold Spring Harbor Laboratory Press, Cold Spring Harbor, NY.

Varki A (2001) Loss of N-glycolylneuraminic acid in humans: Mechanisms, consequences, and implications for hominid evolution. *Am J Phys Anthropol* 33(suppl): 54–69.

Vicarro C, Zhou J, Ober RJ and Ward ES (2005) Engineering the Fc region of immunoglobulin G to modulate *in vivo* antibody levels. *Nat Biotechnol* 23: 283–288. [E-pub 25 September 2005].

Voice KJ and Lachmann PJ (1997) Neutrophil Fc gamma and complement receptors involved in binding soluble IgG immune complexes and in specific granule release induced by soluble IgG immune complexes. Eur J Immunol. 27:2514–23.

Walsh G and Jefferis R (2006) Post-translational modifications in the context of therapeutic proteins. *Nat Biotechnol* 24: 1241–1252.

Wong CH (2005) Protein glycosylation: new challenges and opportunities. *J Org Chem* 70: 4219–4225.

Woof JM and Mestecky J (2005) Mucosal immunoglobulins. *Immunol Rev* 206: 64–68.

Woof JM and Burton DR (2004) Human antibody-Fc receptor interactions illuminated by crystal structures. *Nat Rev Immunol* 4: 89–99.

Wright A and Morrison SL (1998) Effect of C2-associated carbohydrate structure on Ig effector function: studies with chimeric mouse-human IgG1 antibodies in glycosylation mutants of Chinese hamster ovary cells. *J Immunol* 160: 3393–402.

Wright A, Morrison SL and Kobata A (1995) Glycosylation of the variable region of immunoglobulin G – site specific maturation of the sugar chains. *Mol Immunol* 32: 931–940.

Wycoff K (2005) Secretory IgA antibodies from plants. *Curr Pharm Des* 11: 2429–2437.

Yang M and Butler M (2000) Effect of ammonia on the glycosylation of human recombinant erythropoietin in culture. *Biotechnol Prog* 16: 751–759.

Ymane-Ohuki N, Kinoshita S, Inoue-Urakubo M *et al.* (2004) Establishment of FUT8 knockout Chinese hamster ovary cells: an ideal host cell line for producing completely defucosylated antibodies with enhanced antibody-dependent cellular cytotoxicity. *Biotechnol Bioeng* **87**: 614–622.

Youings A, Chang SC, Dwek RA and Scragg IG (1996) Site-specific glycosylation of human immunoglobulin G is altered in four rheumatoid arthritis patients. *Biochem J* **314**: 621–630.

Zhou J, Johnson JE, Ghetie V, Ober RJ and Ward ES (2003) Generation of mutated variants of the human form of the MHC class I-related receptor, FcRn, with increased affinity for mouse immunoglobulin G. *J Mol Biol* **332**: 901–913.

Metabolic engineering to control glycosylation

7

Amy Shen, Domingos Ng, John Joly, Brad Snedecor, Yanmei Lu, Gloria Meng, Gerald Nakamura and Lynne Krummen

7.1 Introduction

Recombinant therapeutic proteins are commonly produced in several mammalian host cell lines including murine myeloma NS0 and Chinese hamster ovary (CHO) cells (Andersen and Krummen, 2002; Chu and Robinson, 2001). Each cell line has advantages and disadvantages in terms of productivity and the characteristics of the proteins produced by the cells. Choices of commercial production cell lines often balance the need for high productivity with the ability to deliver the product quality attributes required of a given product. One important class of therapeutic recombinant proteins which often requires high titer processes is monoclonal antibodies. Some monoclonal antibodies need effector functions, mediated through the Fc region, to elicit their biological functions. An example is rituximab (Rituxan™, Genentech and Biogen-Idec), a chimeric monoclonal antibody which binds to cell surface CD-20 and results in B-cell depletion (Cartron *et al.*, 2002; Idusogie *et al.*, 2000). Other antibodies, such as bevacizumab (Avastin™, Genentech), a humanized anti-VEGF (vascular endothelial growth factor) antibody, rely only on the ability to bind soluble ligand to elicit activity (Wang *et al.*, 2004); Fc effector functions are not required.

Monoclonal antibodies produced in mammalian host cells contain an N-linked glycosylation site at Asn 297 of each heavy chain (two per intact antibody molecule). Glycans on antibodies are typically complex biatennary structures with very low or no bisecting *N*-acetylglucosamine (bisecting GlcNAc) and high levels of core fucosylation (Saba *et al.*, 2002). Glycan termini contain very low or no terminal sialic acid and variable amounts of galactose. Considerable work shows that changes to the sugar composition of the antibody glycan structure can alter Fc effector functions (Kumpel *et al.*, 1994, 1995; Schuster *et al.*, 2005; Shields *et al.*, 2002; Umana *et al.*, 1999). In particular, nonfucosylated structures have recently been associated with dramatically increased *in vitro* antibody-dependent cellular cytotoxicity (ADCC) activity (Shields *et al.*, 2002; Shinkawa *et al.*, 2003). Typically, 90–95% of N-linked glycans found in the Fc region of antibodies produced by the CHO cell lines used for commercial purposes are fucosylated. In contrast, antibodies produced in the YB2/0 cell line (rat myeloma), and the Lec13 cell line, a lectin mutant of CHO line which has a deficient GDP-mannose 4,6-dehydratase leading to the deficiency of GDP-fucose or GDP-sugar intermediates that are the substrate of α-1,6-fucosyltransferase (Ripka *et al.*, 1986), have fewer fucosylated glycans and correspondingly greater effector functions (Shields *et al.*, 2002; Shinkawa *et al.*, 2003). However, the low protein productivity from these cell lines usually makes them unsuitable for manufacturing commercial quantities of therapeutic antibodies. The enzyme responsible for addition of core fucose to N-linked glycan structures is α-1, 6 fucosyltransferase, which is encoded by a gene called FUT8 (Martinez-Duncker *et al.*, 2004; Yamaguchi *et al.*, 2000). Several laboratories, including our own, have successfully employed RNA interference (RNAi) or knock-out techniques to engineer CHO cells to either decrease the FUT8 mRNA transcript levels or knock out gene expression entirely (Mori *et al.*, 2004; Yamane-Ohnuki *et al.*, 2004). This chapter will describe the results we have obtained using a variety of methods to metabolically engineer dihydrofolate reductase (DHFR) negative CHO cells to produce therapeutic antibodies with low fucose content using RNAi technology.

7.2 Manipulation of fucose content using RNAi technology in CHO cells

7.2.1 Metabolic engineering of fucose content with an existing antibody production line

Cloning of FUT8 cDNA and Flag tagging of isolated cDNA

In order to achieve nonfucosylated antibodies in CHO cells, an RNAi approach was employed to knock down the expression of the endogenous FUT8 gene. A 3' 0.98-kb fragment of the FUT8 coding sequence (Genbank accession no. P_AAC63891) was cloned by reverse transcription polymerase chain reaction (RT-PCR) using total RNA purified from CHO cells. An eight amino acid Flag tag (MetAspTyrLysAspAspAspAspLys) was added to the 5' end of the isolated partial cDNA sequence.

Small inhibitory RNA (siRNA) probe design and cloning into the expression vector

The method we used to design siRNA probes to target the CHO FUT8 gene was described by Elbashir *et al.* (2002). Five siRNA probes were constructed using annealed synthetic oligonucleotides independently cloned into the pSilencer 3.1-H1 hygro plasmid (*Figure 7.1*) from Ambion, Inc. (Austin, TX) to produce short hairpin siRNAs. Each siRNA probe consisted of a 19-nucleotide sense sequence specific to the FUT8 gene, linked to its reverse complement antisense sequence by a nine-nucleotide hairpin-loop sequence (*Figure 7.2*). The siRNA probe 1 (RNAi1), does not target sequences contained in the Flag-tagged FUT8 gene fragment but does target sequences in the endogenous FUT8 gene (*Figure 7.2*). The ability of these probes to cleave the FUT8 transcript was tested by transient cotransfection of each siRNA expression plasmid with the Flag-tagged FUT8 plasmids into CHO cells. Cells were lysed 24 h after transfection and the cell lysate was analyzed by western blot with anti-FlagM2 antibody (Sigma, MO). RNAi1 transfected cells, as expected, showed strong expression of the Flag-tagged FUT8 product since the Flag-tagged FUT8 fusion protein does not contain the sequence targeted by this probe (*Figure 7.3*). In contrast, siRNA probes 2

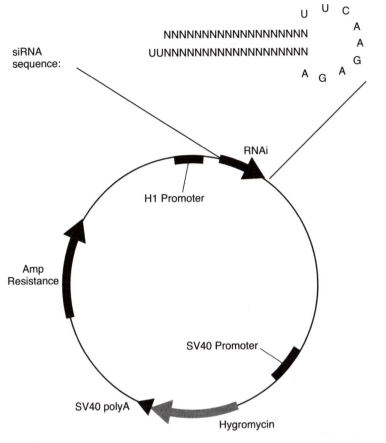

Figure 7.1

Plasmid vector used to express small inhibitory RNA. The vector was purchased from Ambion (Austin, TX). The DNA sequences encoding siRNA probes were cloned into BamHI and HindIII sites under the control of Pol III type H1 promoter. The transcript from H1 promoter forms a hairpin-loop siRNA.

(RNAi2) through 5 all have various degrees of inhibitory effects on Flag-tagged FUT8 fusion protein expression. RNAi2 and RNAi4, which demonstrated markedly stronger inhibitory effects than RNAi3 and RNAi5, were chosen for further evaluation.

Fucose content of stably expressed antibodies manipulated by transient siRNA expression

RNAi2 and 4 plasmids were transiently transfected into a previously established anti-CD20 antibody secreting CHO stable cell line (antibody A). The transfected cells were then separately seeded into 250-mL spinner vessels in serum-free medium for antibody production. The secreted antibody in the harvested cell culture fluid (HCCF) was purified by protein A column chromatography and N-linked oligosaccharides were analyzed for fucose content by matrix-assisted laser desorption/ionization time-of-flight mass

(A)

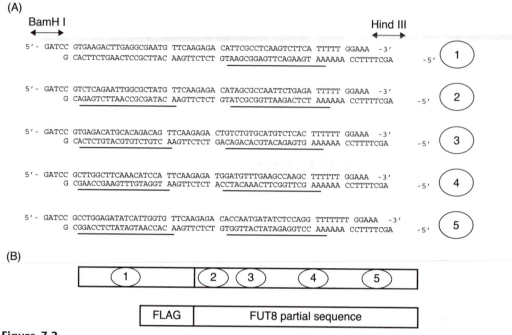

(B)

Figure 7.2

SiRNA probe sequences and their relative positions in full length FUT8 and FLAG-tagged partial FUT8 genes. Each siRNA probe sequence is underlined (A). The underlined sequence close to BamHI site is complementary to the FUT8 mRNA sequence. The two underlined sequences are complementary to each other resulting in formation of the hairpin loop siRNA. The siRNA targeting regions in FUT8 gene product are schematically indicated (B). The siRNA probe 1 targets the 5¢ end of endogenous FUT8 gene product, but has no complementary targeting sequence in the FLAG-tagged partial FUT8 construct.

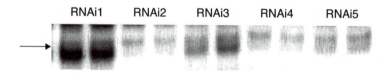

Figure 7.3

Western blot analysis of lysates from the cotransfection of the individual siRNA probes and the FLAG-tagged partial FUT8 construct. Five individual siRNA expression constructs were transiently cotransfected with FLAG-tagged partial FUT8 construct. Cell lysates containing equal amounts of cellular proteins were analyzed by western blot with anti-FLAG antibody (Sigma, MO). RNAi1 can not downregulate FLAG-tagged partial FUT8 expression and therefore served as a negative control. The arrow points to the FLAG-tagged partial FUT8 protein.

spectroscopy (MALDI-TOF) as described (Papac *et al.*, 1998). As shown in *Table 7.1*, the antibody molecules produced from the cells transiently transfected by the RNAi2 and RNAi4 plasmids contained 42% and 40% nonfucosylated glycan, respectively, whereas antibodies produced from mock transfected cells had about 3% nonfucosylated glycan, a level typical of antibodies generated from CHO cells. Glycans isolated from antibodies

Table 7.1 Glycan compositions of antibodies produced from RNAi construct transfected stable antibody production cell line

Sample	%G0	%G1	%G2	%–F
Antibody A control	71	26	3	3
Antibody A.RNAi2	73	24	3	42
Antibody A.RNAi4	65	29	6	40

produced from both RNAi plasmid transfected and mock-transfected cells had similar distributions of galactose content when structures with no galactose (G0), one galactose (G1), and two galactose (G2) were compared. These data show that the fucose content of antibodies secreted from a stable production cell line can be decreased by transient RNAi plasmid transfection and that the effect does not alter the other main glycan compositions including G0, G1, and G2 distributions. The observation that RNAi2 and RNAi4 plasmids had similar efficacy in decreasing endogenous FUT8 gene expression may indicate that there is a limitation to the overall effectiveness of this strategy. To determine whether transfection efficiency was a limiting factor, stable transfection of the RNAi4 plasmid was examined.

To construct a CHO-producing cell line that stably produces antibody A molecules with low levels of fucosylation, the antibody production cell line was transfected with RNAi4 plasmid and stable transfectants were selected by exposure to 500 µg mL^{-1} hygromycin. Hygromycin positive clones were seeded in a 96-well tissue culture plate and screened by Taqman for endogenous FUT8 mRNA expression. As shown in *Figure 7.4*, clones showed different levels of FUT8 gene expression. This was expected since FUT8 gene expression would be expected to be inversely related to the expression of siRNA resulting from integrated plasmid sequences which can vary in a position dependent manner. Two clones (clones 5F and 7C) showing low

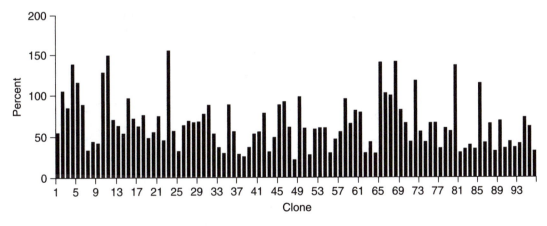

Figure 7.4

Taqman analysis of endogenous FUT8 gene expression by clones stably transfected with the RNAi4 construct. Total RNAs were isolated from individual clones which were stably transfected with RNAi4 construct. The relative FUT8 expression is normalized to the level seen in the mock-transfected antibody production cell line. The data represent the mean of two replicate results.

levels of FUT8 mRNA expression in Taqman assay were scaled up to produce antibodies in 250-mL spinner vessels. Antibodies in the HCCF were purified by protein A column chromatography and N-glycans were analyzed for fucose content by MALDI-TOF. *Table 7.2* shows that 65% and 70% of the glycans produced by these clones were not fucosylated. To investigate the stability of the inhibition of FUT8 gene expression, both stably transfected clones were cultured in 250-mL spinner vessels and passaged every 3 to 4 days. At different time intervals, secreted antibodies were purified by protein A column chromatography and analyzed for fucose content. As shown in *Figure 7.5*, the level of antibody non-fucosylation for both clones was stable over the 79-day period studied. At the end of 79-day seed train duration, clone 7C was further evaluated in a 10-L bioreactor. Again around 75% of glycan structures were found to lack fucose following scale-up (*Figure 7.6*). Antibody titers as well as the percent of G0, G1, and G2 on antibody glycans were also in the expected range at the end of the bioreactor run (data not shown). Therefore, re-transfection of RNAi plasmid into an established production cell line provides a viable approach to generating production cell lines capable of stable production of commercial amounts of a therapeutic antibody with controlled amounts of glycan fucosylation.

Table 7.2 Glycan compositions of two stable clones transfected with RNAi4 construct

Clone	%G0	%G1	%G2	%–F
5F	65	33	2	65
7C	66	32	2	70

7.2.2. Metabolic engineering of fucose content with simultaneous new stable cell line generation

The process illustrated above represents a two-step approach, requiring the existence of a stable antibody producing cell line before RNAi plasmid

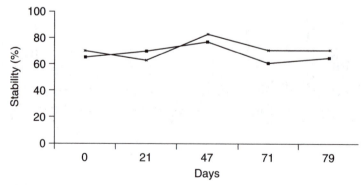

Figure 7.5

Stability of non-fucosylation level. Two stable clones, 5F and 7C, were cultured in 250-mL spinner vessels and passaged on a day 3 or day 4 schedule. Samples of secreted antibodies were taken at different times and analyzed for fucosylation levels. Solid square, clone 5F; solid triangle, clone 7C.

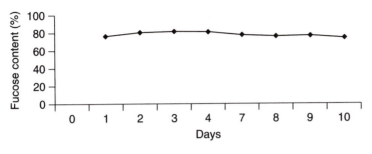

Figure 7.6

Evaluation of the non-fucosylated of antibody glycan content produced from clone 7C in a bioreactor. Clone 7C was evaluated in a 10-L bioreactor vessel for 10 days. The samples were collected daily and analyzed by MALDI-TOF for fucose content after protein A column chromatography purification. Fucose contents vary within a narrow range centered around 75% during the 10-day bioreactor fermentation.

transfection. To shorten the time needed for this process, a one-step approach was explored wherein the siRNA cassette(s) was included on the expression plasmid encoding the therapeutic protein of interest. Two plasmids used in this new approach are illustrated in *Figure 7.7*. Two constructs were developed containing one siRNA unit, RNAi2, or two siRNA units, RNAi2 and RNAi4. The rationale for incorporating the second siRNA cassette onto the expression plasmid was that targeting the endogenous CHO FUT8 gene transcript at two different places within the open reading frame (ORF) might lead to more complete cleavage of the FUT8 gene transcript. The antibody molecule (antibody B) used for the study was a variant of the anti-CD20 monoclonal antibody used for previous experiments. This antibody contains a few amino acid changes in the Fc region designed to enhance ADCC activity. The expression plasmids containing the antibody expression and siRNA cassettes were first tested to see if antibody and siRNA could be expressed simultaneously in a transient transfection. The antibody expressed from each of the two plasmids was purified by protein A column chromatography and glycans were analyzed for fucose levels. As shown in *Figure 7.8*, only 9% of the antibody glycans from cultures transfected with control plasmids containing

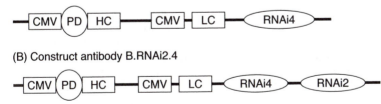

(A) Construct antibody B.RNAi4

(B) Construct antibody B.RNAi2.4

Figure 7.7

Expression construct containing antibody transcription unit and RNAi unit. PD denotes puromycin DHFR fusion gene located in the intron of heavy chain transcription unit. Antibody B.RNAi4 plasmid contains one RNAi4 unit located 3′ to the light chain transcription unit. Antibody B.RNAi2.4 plasmid contains both RNAi2 and RNAi4 units located 3′ to the light chain transcription unit.

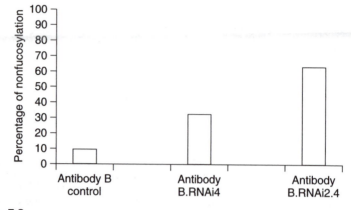

Figure 7.8

Non-fucosylation levels of antibodies from cells transiently transfected with the constructs containing both an antibody expression unit and RNAi unit on the same plasmid. The control plasmid of antibody B contains antibody expression unit only.

no siRNA cassette lacked fucose, whereas 33% and 65% of the antibody glycans isolated from the cells transfected with one or two siRNA cassettes respectively were nonfucosylated. These data indicate that under transient transfection conditions, incorporating two siRNA cassettes onto the expression plasmid leads to the production of antibodies with lower fucosylation than observed when the plasmid contained one siRNA unit, suggesting an additive effect of the two siRNA transcripts.

The same two antibody-expressing plasmids were then used to create stable antibody producing cell lines. The plasmids were separately transfected into blank CHO cells in 250-mL spinner vessels and exposed to 25 nM methotrexate (MTX). From each transfection, 72 clones were seeded into a 96-well plate and screened for antibody expression. About 20% of clones with high antibody expression were analyzed for FUT8 mRNA expression by Taqman. As shown in *Figure 7.9*, the clones from the expression plasmid containing two siRNA cassettes had generally lower FUT8 mRNA levels compared to the clones from the plasmid containing one siRNA transcription unit. Six clones with the lowest FUT8 mRNA expression levels were selected for further evaluation of antibody productivity and fucose content along with the most productive clone from the control transfection. Equal numbers of cells from each clone were seeded into a productivity assay. Expression titers are shown in *Figure 7.10*. Although a trend towards decreased titers was noted in clones resulting from the transfection of plasmids with two versus zero or one siRNA unit, the best titers from each case were similar. This suggests that the addition of multiple siRNA transcription unit(s) on the antibody expression construct does not have a consistently detrimental effect on antibody expression. Fucose contents of antibodies produced by the control and transfected clones were studied using cultures maintained in 250-mL spinner vessels. As shown in *Figure 7.11*, glycans produced in these cultures had a range of fucosylation. In the control case, only about 3% of antibody glycans were not fucosylated, whereas glycans produced by the clones transfected with RNAi plasmids had markedly lower levels of fucose. Although relatively small

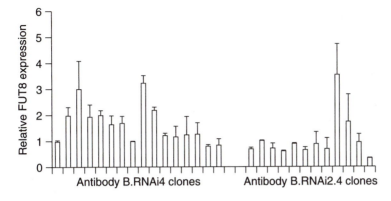

Figure 7.9

Analysis of relative endogenous FUT8 gene expression by Taqman. Clones from antibody B.RNAi4 and antibody B.RNAi2.4 stable transfection were analyzed for endogenous FUT8 mRNA expression by Taqman. The relative expression levels of all clones were normalized to one arbitrarily chosen clone. The data represent the mean of two replicate results.

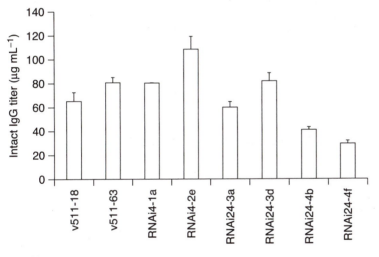

Figure 7.10

Antibody expression of stable clones with lower FUT8 gene expression. Two antibody B.RNAi clones and four antibody B.RNAi2.4 clones with lower FUT8 gene expression in Taqman assay were evaluated for the intact antibody production by ELISA. The best expressor, clone 63, from control plasmid transfection was also included as control. The data represent the mean of triplicate results.

numbers of clones were evaluated, clones with two siRNA transcription units on the expression plasmid generally tended to have lower fucosylation levels than clones with one siRNA transcription unit. In clones transfected with two RNAi units, we observed that up to 95% of glycans were not fucosylated. Similar results were confirmed when this clone was cultured in a 2-L bioreactor (data not shown). These data demonstrate that a one-step approach can be

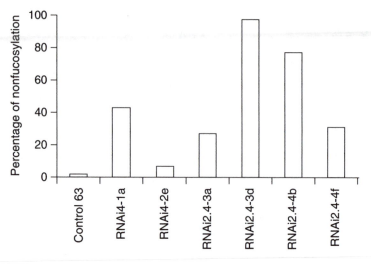

Figure 7.11

Non-fucosylation level analysis. Antibodies produced from six clones, RNAi4 1a, 2e, RNAi2.4 3a, 3d, 4b and 4f, in 250-mL spinner vessels were protein A column purified and were assayed in MALDI-TOF for non-fucosylation level. In this case, the best expressor from control plasmid of antibody B was included as a control.

taken to construct cell lines that exhibit high antibody productivity and controlled levels of fucosylation.

7.2.3 Effect of fucosylation levels on FcγR binding

FcγR binding to anti-CD20 antibodies with different fucose contents

There are three groups of human Fcγ receptors: FcγRI, FcγRII, and FcγRIII. Some of these have a functional allelic polymorphism generating allotypes with different receptor properties (Dijstelbloem *et al.*, 1999; Lehrnbecher *et al.*, 1999). FcγRIIIa(F158) has phenylalanine at position 158 and has a lower binding affinity for the Fc region of human IgGs than FcγRIIIa(V158) which has a valine at position 158 (Shields *et al.*, 2001, 2002). To see the effect of fucose content of antibody A on Fcγ binding, antibodies containing 3% or 40–42% nonfucosylated glycans were assessed for their binding abilities to FcγRI, II and III. As shown (*Figure 7.12*), there is no appreciable difference among the antibodies for binding to FgRI, FcγRIIA or FcγRIIB receptors. However, there is a marked improvement in binding affinity to FcγRIIIa receptors for antibodies containing less fucose compared to control antibodies. Due to the lack of binding difference for FcγRI, FcγRIIA, and FcγRIIB in our ELISA assays, additional studies focused on the relationship of different levels of fucosylation to FcγRIIIa binding activity. Antibodies containing five different levels of fucosylation were assayed for their binding to FcγRIIIa. The fold increases in affinity for FcγRIIIa(F158) and

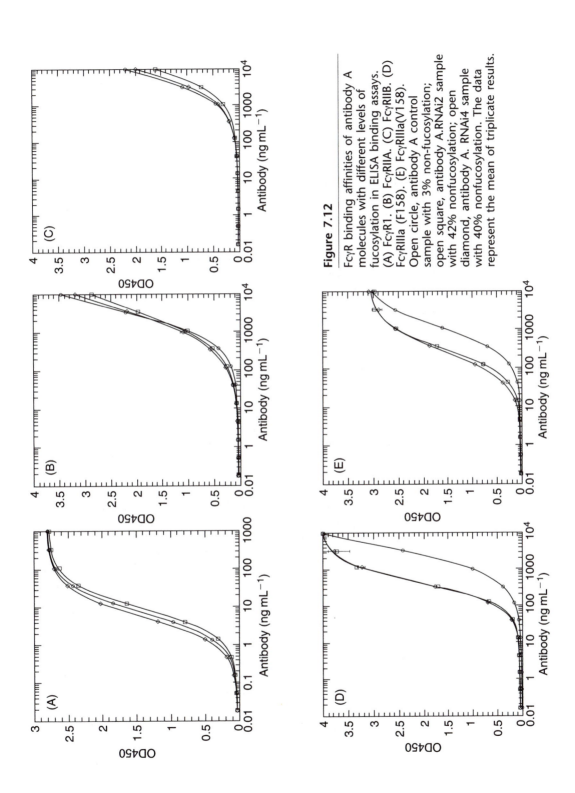

Figure 7.12

Fcγr binding affinities of antibody A molecules with different levels of fucosylation in ELISA binding assays. (A) FcγR1. (B) FcγRIIA. (C) FcγRIIB. (D) FcγRIIIa (F158). (E) FcγRIIIa(V158). Open circle, antibody A control sample with 3% non-fucosylation; open square, antibody A.RNAi2 sample with 42% nonfucosylation; open diamond, antibody A. RNAi4 sample with 40% nonfucosylation. The data represent the mean of triplicate results.

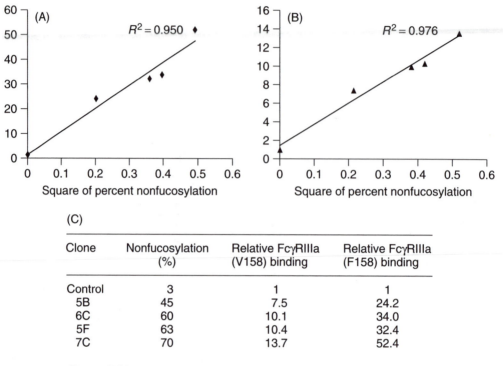

Figure 7.13

Relationship between FcγRIIIa binding and non-fucosylation level for antibody A molecules in ELISA assay. (A) Relative fold increase in affinity to FcγRIIIa (F158) plotted against the square of percentage of non-fucosylation; (B) relative fold increase in affinity to FcγRIIIa(V158) plotted against the square of percentage of nonfucosylation; (C) FcγRIIIa binding affinities versus nonfucosylation contents.

FcγRIIIa(V158) are shown in *Figure 7.13* for each antibody tested. When the fold increase was plotted against the square of the percentage of nonfucosylated material in each antibody sample, a linear relationship was seen for both FcγRIIIa variants. Intact human IgG1 contains two heavy chains, each with a N-glycosylation site at Asn297 in the CH2 domain of the Fc region. Therefore there are three possibilities for the Fc in terms of fucose occupancy of the core carbohydrate structure: either one heavy chain is fucosylated and one is not, or both heavy chains are fucosylated, or neither heavy chain is fucosylated. The linear relationship between the fold increase in affinity to FcγRIIIa and the square of the percentage of nonfucosylated glycans indicates that in this case antibody molecules with neither heavy chain fucosylated may provide the major contribution to the improvement of increased FcγRIIIa binding affinity.

FcγRIII binding to anti-CD20 antibodies with modified primary sequences and different fucose contents

Another way to improve the binding affinity of an antibody to FcγRIIIa is to change the amino acid sequence(s) in the Fc region. Shields *et al.* (2002)

Table 7.3 FcγRIIIa binding affinities of antibody molecules with different nonfucosylation levels in ELISA binding assay

Sample	Percent nonfucosylation	Relative FCγRIIIa (V158) binding	Relative FCγRIIIa (F158) binding
Antibody A control	3	1	1
Antibody A clone 5F	65	10.4	32.4
Antibody B control	3	10.8	37.0
Antibody B.RNAi2.4	65	51.9	217.2
Antibody B.RNAi2.4	95	72.9	328.8

have successfully applied this approach to improve binding affinity to FcγRIIIa for a variety of antibody molecules. Using the CHO Lec 13 cell line to produce antibodies with lower fucose content, they have also shown that when the antibody molecule contains both enhancing amino acid changes in the Fc region and lower fucose content, there is an additive effect on affinity for FcγRIIIa. In the present studies, we compared the effects of altering the fucose content of the antibody (antibody A) which has a wild-type Fc to that seen when the same carbohydrate changes were made to the variant of the anti-CD20 molecule (antibody B) that contains a few amino acid sequence changes in the Fc region to improve FcγRIIIa binding. Interestingly, we found that reducing the fucose content of the antibody B molecule had a less dramatic impact on FcγRIIIa binding than when similar fucose modifications were made to the antibody A molecule (*Table 7.3*). Whether this is molecule dependent or this is because binding to FcγRIIIa is saturable and antibody B molecules approach the maximum binding capability more closely than antibody A so that the incremental enhancement of lowering fucose is diminished is under investigation.

7.2.4 Effects of fucose content on antibody-dependent cellular cytotoxicity

To compare the impact of lowering the fucose contents of the two anti-CD20 molecules on their ability to elicit antibody-dependent cellular cytotoxicity (ADCC), antibody samples containing various levels of nonfucosylated glycans were assayed using natural killer (NK) cells as effector cells (Shields *et al.*, 2002). Representative data from one of several experiments is shown in *Figure 7.14*. Compared with the control molecules with ~3% nonfucosylated glycan, antibody A and B molecules with 65–70% nonfucosylated glycans showed markedly increased ADCC activities (*Figure 7.14A*). As expected, the antibody B molecules with normal levels of fucosylation had significantly higher ADCC activity than the normally fucosylated antibody A (*Figure 7.14B*).

7.3 Discussion

The composition of the carbohydrate structure attached to Asn 297 of recombinant antibodies has been shown to affect the binding affinity of antibodies to three classes of FcγRs including FcgI, FcgII, and FcgIII. FcγRs link antibody-mediated immune responses with cellular effector functions

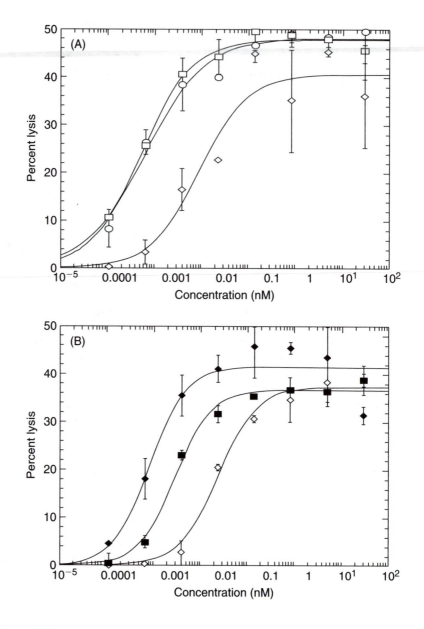

Figure 7.14

Evaluation of ADCC activities of antibodies with different nonfucosylation contents. The ADCC assays were performed with natural killer (NK) cells as described (Shields et al., 2002). (A) Representative plot for FcγRIIIa(V158/F158) donor; (B) representative plot for FcγRIIIa(V158/V158) donor. Open diamond, antibody A control with 3% nonfucosylation; open oval, antibody A with 70% nonfucosylation; open square, antibody A with 65% nonfucosylation; solid square, antibody B control with 3% nonfucosylation; solid diamond, antibody B with 65% nonfucosylation.

including ADCC (Cartron *et al.*, 2002; Davis *et al.*, 2001). The recombinant antibodies produced from nonengineered CHO cells usually contain asialyl, biantennary N-linked glycan structures that terminate with variable amounts of galactose (reviewed by Jefferis, 2005). The core composition of antibody glycans is characterized as highly fucosylated (Saba *et al.*, 2002; Shields *et al.*, 2002) but without appreciable amounts of bisecting GlcNAc. The effect of changes in galactosylation on ADCC activity have been previously studied (Boyd *et al.*, 1995; Kumpel *et al.*, 1994). In some cases no appreciable effect of galactose levels on ADCC activity was reported (Boyd *et al.*, 1995), while others report a slight enhancement of ADCC activity (Kumpel *et al.*, 1994; Kumpel *et al.*, 1995). The effect of bisecting GlcNAc on ADCC has also been studied by several groups. Antibodies with bisecting GlcNAc content ranging from about 20% to 75% have been prepared by expression in either YB2/0 cells or CHO cells engineered to overexpress *N*-acetylglucosaminyltransferase III (GnTIII) (Davis *et al.*, 2001; Schuster *et al.*, 2005; Shinkawa *et al.*, 2003). A common observation in these studies is that the magnitude of the improvement of ADCC activity does not correlate well with the change in amount of bisecting GlcNAc structure (Shinkawa *et al.*, 2003). It is possible that the relative enhancement in ADCC from bisecting GlcNAc manipulation might be molecule dependent. However, another explanation, which is supported by recent studies (Shinkawa *et al.*, 2003) is that manipulation of the bisecting GlcNAc content of the glycan leads to concomitant alterations in the core fucosylation level that overestimate the effect of bisecting GlcNAc content on ADCC. When the two effects were separated, it was found that compared with nonfucosylation, the effect of bisecting GlcNAc on ADCC was minimal. Only addition of high levels of bisecting GlcNAc had clear effects on ADCC (Shinkawa *et al.*, 2003). Therefore, fucosylation levels appear to have the most significant effect on ADCC activity (Niwa *et al.*, 2004b, 2005; Shinkawa *et al.*, 2003). Okazaki *et al.* (2004) have proposed that when fucose is depleted from glycans on antibody molecules, the binding enthalpy and association rate between IgG1 and FcγRIIIa is enhanced. To make antibodies with lower fucosylation, Shields *et al.* (2002) used a lectin mutant CHO Lec13 cell line. Unfortunately, this cell line is not robust enough to be an industrial host, since the productivity is generally about 100 times lower than wild-type CHO cells in our hands.

In recent years, RNA interference approaches have been used successfully to knock down gene expression (Dykxhoorn *et al.*, 2003; Elbashir *et al.*, 2002; Irie *et al.*, 2002; Katome *et al.*, 2003; Khanzada *et al.*, 2006; Sui *et al.*, 2002; Taniguchi et al 2004). Mori *et al.* (2004) transfected siRNA plasmids targeting the FUT8 messenger RNA into a stable antibody producing cell line and demonstrated production of antibodies that contained up to 70% nonfucosylated glycan. In the present study we saw similar results when we transfected a siRNA construct employing the H1 promoter driving a different, single FUT-8 siRNA unit into a stable antibody secreting cell line. Our data also show that expression of siRNA activity is stable for at least 79 days in spinner culture. The two clones studied had normal growth rate and maintained fucosylation within a narrow range throughout this period. Furthermore, it appears that no differences in glycosylation other than fucose content result from this RNAi approach, and no adverse impact on net antibody productivity was observed. Importantly, the altered fucose content was also maintained even after scale-up in a 10-L fed-batch bioreactor culture.

In this chapter, we also described a novel streamlined way to metabolically engineer CHO cells to produce even more highly (as high as 95%) nonfucosylated antibodies by incorporating the antibody heavy chain and light chain transcription units along with one to two siRNA transcription unit(s) onto the same plasmid. The two siRNA transcripts used in this approach target different coding regions in the FUT8 gene and are directed by separate Pol III type promoters, H1 and U6. The data from Taqman analyses show that in general, the clones from the construct containing two RNAi units have lower fucosylation than the clones derived from the construct containing one RNAi unit. It is reasonable to speculate that there is a maximum siRNA message level transcribed from the H1 promoter, therefore, there are more siRNA molecules available to cleave FUT8 message when there is another RNAi unit present. Maximizing the content of nonfucosylated heavy chains may be critical in obtaining the full benefit of lowered fucosylation. In our studies we observed that FcγRIIIa binding activity increased approximately with the square of the percentage of nonfucosylated glycans. This indicates that antibody molecules with no fucose on either Fc region may have the strongest contribution to increased FcγRIIIa binding activity, suggesting that a FUT8 knock out CHO cell line (Yamane-Ohnuki *et al.*, 2004) may be an optimal host cell line to make the therapeutic proteins which need effector functions to elicit their biological functions.

In addition, reducing fucosylation also appeared to have a great impact on increasing IgG1 binding affinity in ELISA assay for the low affinity receptor FcγRIIIa(F158) than the high affinity receptor FcγRIIIa(V158). FcγRIIIa(F158) allele is more common than FcγRIIIa(V158) allele in normal African Americans and Caucasians, and about 85–90% of the whole human population carries a copy of the FcγRIIIa(F158) allele (Lehrnbecher *et al.*, 1999). Taken together these data suggest that engineering the fucose content of recombinant antibodies may be a clinically important means to enhance the activity of therapeutic antibodies which rely on FcγRIII-mediated effector function for activity.

Acknowledgments

The authors thank Henry Lowman for valuable discussion; Dot Reilly and George Dutina for performing bioreactor fermentations; and Rod Keck for glycan assays.

References

Andersen DC and Krummen L (2002) Recombinant protein expression for therapeutic applications. *Curr Opin Biotech* **13**: 117–123.

Boyd PN, Lines AC and Patel AK (1995) The effect of the removal of sialic acid, galactose and total carbohydrate on the functional activity of campath-1H. *Mol Immunol* **32**: 1311–1318.

Cartron G, Dacheux L, Salles G, Solal-Celigny P, Bardos P, Colombat P and Watier H (2002) Therapeutic activity of humanized anti-CD20 monoclonal antibody and polymorphism in IgG Fc receptor FcγRIIIa gene. *Blood* **99**: 754–758.

Chu L and Robinson DK (2001) Industrial choices for protein production by large-scale cell culture. *Curr Opin Biotechnol* **12**: 180–187.

Davis J, Jiang L, LaBarre MJ, Andersen D and Reff M (2001) Expression of GnTIII in a recombinant anti-CD20 CHO production cell line: expression of antibodies of altered glycoforms leads to an increase in ADCC through higher affinity for FcRIII. *Biotechnol Bioeng* **74**: 288–294.

Dijstelbloem HM, Scheepers RH, Oosh WW *et al* (1999) Fcγ receptor polymorphisms in Wegener's granulomatosis: risk factors for disease relapse. *Arthritis Rheum* **42**: 1823–1827.

Dykxhoorn DM, Novina CD and Sharp PA (2003) Killing the messenger: short RNAs that silence gene expression. *Mol Cell Biol* **4**: 457–467.

Elbashir SM, Harborth J, Weber K and Tuschl T (2002) Analysis of gene function in somatic mammalian cells using small interfering RNAs. *Methods* **26**: 199–213.

Idusogie EE, Presta LG, Gazzano-Santoro H, Totpal K, Wong PY, Ultsch M, Meng YG and Mulkerrin MG (2000) Mapping of the C1q binding site on Rituxan, a chimeric antibody with a human IgG1 Fc. *J Immunol* **164**: 4178–4184.

Irie N, Sakai N, Ueyama T, Kajimoto T, Shirai Y and Saito N (2002) Subtype- and species-specific knockdown of PKC using short interfering RNA. *Biochem Biophys Res Commun* **298**: 738–743.

Jefferis R (2005) Glycosylation of recombinant antibody therapeutics. *Biotechnol Prog* **21**: 11–16.

Katome T, Obata T, Matsushima R, Masuyama N, Cantley LC, Gotoh Y, Kishi K, Shiota H and Ebina Y (2003) Use of RNA interference-mediated gene silencing and adenoviral overexpression to elucidate the roles of AKT/protein kinase B isoforms in insulin actions. *J Biol Chem* **278**: 28312–28323.

Khanzada UK, Pardo OE, Meier C, Downward J, Seckl MJ and Arcaro A (2006) Potent inhibition of small-cell lung cancer cell growth by simvastatin reveals selective functions of Ras isoforms in growth factor signalling. *Oncogene* **25**:877–887.

Kumpel BM, Rademacher TW, Rook GA, Williams PJ and Wilson IB (1994) Galacosylation of human IgG monoclonal anti-D produced by EBV-transformed B-lymphoblastoid cell lines is dependent on culture method and affects Fc receptor-mediated functional activity. *Hum Antib Hybrid* **5**: 143–151.

Kumpel BM, Wang Y, Griffiths HL, Hadley AG and Rook GA (1995) The biological activity of human monoclonal IgG anti-D is reduced by beta-galactosidase treatment. *Hum Antib Hybrid* **6**: 82–88.

Lehrnbecher T, Foster CB, Zhu S, Leitman SF, Goldin LR, Huppi K and Chanock SJ (1999) Variant genotypes of the low-affinity Fcγ receptors in two control populations and a review of low-affinity Fcγ receptor polymorphisms in control and disease populations. *Blood* **94**: 4220–4232.

Martinez-Duncker IM, Michalski JC, Bauvy C, Candelier JJ, Mennesson B, Codogno P, Oriol R and Mollicone R (2004) Activity and tissue distribution of splice variants of α1,6-fucosyltransferase in human embryogenesis. *Glycobiology* **14**: 13–25.

Mori K, Kamochi RK, Ohnuki NY *et al.* (2004) Engineering Chinese hamster ovary cells to maximize effector function of produced antibodies using FUT8 siRNA. *Biotechnol Bioeng* **88**: 901–908.

Niwa R, Hatanaka S, Shoji-Hosaka E, Sakurada M, Kobayashi Y, Uehara A, Yokoi H, Nakamura K and Shitara K (2004a) Enhancement of the antibody-dependent cellular cytotoxicity of low-fucose IgG1 is independent of FcRIIIa functional polymorphism. *Clin Cancer Res* **10**: 6248–6255.

Niwa R, Hosaka ES, Sakurada M *et al.* (2004b) Defucosylated chimeric anti-CC chemokine receptor 4 IgG1 with enhanced antibody-dependent cellular cytotoxicity shows potent therapeutic activity to T-cell leukemia and lymphoma. *Cancer Res* **64**: 2127–2133.

Niwa R, Sakurada M, Kobayashi Y, Uehara A, Matsushima K, Ueda R, Nakamura K and Shitara K (2005) Enhanced natural killer cell binding and activation by low-fucose IgG1 antibody results in potent antibody-dependent cellular cytotoxicity induction at lower antigen density. *Clin Cancer Res* **11**: 2327–2336.

Okazaki A, Hosaka ES, Nakamura K, Wakitani M, Uchida K, Kakita S, Tsumoto K, Kumagai I and Shitara K (2004) Fucose depletion from human IgG1 oligosaccharide enhances binding enthalpy and association rate between IgG1 and FcγRIIIa. *J Mol Biol* **336**: 1239–1249.

Papac DJ, Briggs JB, Chin ET and Jones AJS (1998) A high-throughput microscale method to release N-linked oligosaccharides from glycoproteins for matrix-assisted laser desorption/ionization time-of-flight mass spectrometric analysis. *Glycobiology* **8**: 445–454.

Ripka J, Adamany A and Stanley P (1986) Two Chinese hamster ovary glycosylation mutants affected in the conversion of GDP-mannose to GDP-fucose. *Arch Biochem Biophys* **249**: 533–545.

Saba JA, Kunkel JP, Jan DCH, Ens WE, Standing KG, Butler M, Jamieson JC and Perreault H (2002) A study of immunoglobulin G glycosylation in monoclonal and polyclonal species by electrospray and matrix-assisted laser desorption/ionization mass spectrometry. *Anal Biochem* **305**: 16–31.

Schuster M, Umana P, Ferrara C *et al.* (2005) Improved effector functions of a therapeutic monoclonal Lewis Y-specific antibody by glycoform engineering. *Cancer Res* **65**: 7934–7941.

Shields RL, Lai J, Keck R, O'Connell LY, Hong K, Meng YG, Weikert SHA and Presta LG (2002) Lack of fucose on human IgG1 N-linked oligosaccharide improves binding to human Fc γRIII and antibody-dependent cellular toxicity. *J Biol Chem* **277**: 26733–26740.

Shields RL, Lai J, Keck R, O'Connell LY, Hong K, Meng YG, Weikert SH and Presta LG (2001) High-resolution mapping of the binding site on human IgG1 for FcRI, FcRII, FcRIII, and FcRn and design of IgG1 variants with improved binding to the FcR. *J Biol Chem* **276**: 6591–6604.

Shinkawa T, Nakamura K, Yamane N *et al.* (2003) The absence of fucose but not the presence of galactose or bisecting N-acetylglucosamine of human IgG1 complex-type oligosaccharides shows the critical role of enhancing antibody-dependent cellular cytotoxicity. *J Biol Chem* **278**: 3466–3473.

Sui G, Soohoo C, Affar EB, Gay F, Shi Y, Forrester WC and Shi Y (2002) A DNA vector-based RNAi technology to suppress gene expression in mammalian cells. *Proc Natl Acad Sci USA* **99**: 5515–5520.

Taniguchi N, Gu J, Takahashi M and Miyoshi E (2004) Functional glycomics and evidence for gain- and loss-of-functions of target proteins for glycosyltransferase involved in N-glycan biosynthesis: their pivotal roles in growth and development, cancer metastasis and antibody therapy against cancer. *Proc Japan Acad Ser* **B80**: 82–91.

Umana P, Lean-Mairet J, Moudry R, Amstutz H and Bailey JE (1999) Engineered glycoforms of an antineuroblastoma IgG1 with optimized antibody-dependent cellular cytotoxic activity. *Nat Biotechnol* **17**: 176–180.

Wang Y, Fei D, Vanderlaan M and Song A (2004) Biological activity of bevacizumab, a humanized anti-VEGF antibody *in vitro*. *Angiogenesis* **7**: 335–345.

Yamaguchi Y, Ikeda Y, Takahashi T *et al.* (2000) Genomic structure and promoter analysis of the human α1,6-fucosyltransferase gene (FUT8). *Glycobiology* **10**: 637–643.

Yamane-Ohnuki N, Kinoshita S, Urakubo MI *et al.* (2004) Establishment of FUT8 knockout Chinese hamster ovary cells: an ideal host cell line for producing completely defucosylated antibodies with enhanced antibody-dependent cellular cytotoxicity. *Biotechnol Bioeng* **87**: 614–622.

An alternative approach: Humanization of N-glycosylation pathways in yeast

8

Stefan Wildt and Thomas Potgieter

8.1 Introduction

Recombinant DNA technology paved the way for scientific discoveries that have led to substantial contributions in sustaining and improving human health. In the biopharmaceutical industry, genetic engineering tools allowed the production of recombinant therapeutic proteins (previously derived from human sources), alleviating the inherent risks associated with human-derived materials such as batch-to-batch variability and the potential for viral contaminations. In 2003, the global market for biopharmaceuticals was more than $30 billion (Walsh, 2003). Today, biologics are the largest class of new chemical entities being developed (Walsh, 2003). Several hundred therapeutic proteins are in varying stages of clinical trials, with an even larger number in preclinical development.

Therapeutic proteins can be divided into two classes, those without post-translational modifications and those that require post-translational modifications (in most cases N-linked glycosylation) to achieve full biological function. Nonglycosylated proteins or proteins that do not require glycosylation are typically expressed in microbial production systems, such as *Escherichia coli* or fungi like the yeast *Saccharomyces cerevisiae*.

N-linked glycosylation influences intracellular trafficking of glycoproteins, impacts protein folding, determines protein half-life, has been shown to support anchoring of macromolecules to the cell membrane, and is involved in signal transduction and cell–cell interactions (Coloma *et al.*, 2000; Helenius and Aebi, 2001). It is therefore clear that N-linked glycosylation can in most cases be considered as essential for proper functioning of the therapeutic protein.

Although there are a few rare examples of prokaryotes capable of post-translational glycosylation (Thibault, 2001), this function is generally believed to originate in the endoplasmic reticulum (ER) of eukaryotic cells. As a result, eukaryotes are chosen as the preferred hosts for the expression of recombinant proteins requiring N-linked glycosylation. While both fungi and mammals attach a specific oligosaccharide to asparagine in the sequence Asn-X-Ser/Thr/Cys (X represents any amino acid except proline), subsequent processing of the transferred glycan differs significantly between mammalian and fungal cells. A more detailed overview of the process of N-linked glycosylation in yeast and mammals will be provided below. In brief, the assembly of a lipid-linked oligosaccharide $Glc_3Man_9GlcNAc_2$ followed by the transfer to the nascent protein and the removal of three glucose residues and one mannose sugar to yield $Man_8GlcNAc_2$, is conserved between eukaryotes. The biosynthetic glycosylation pathways diverge between yeast and mammals once a glycoprotein leaves the ER and is shuttled through the Golgi.

Yeast and other fungi typically produce high mannose-type N-glycans by adding up to 100 mannose sugars, including beta-linked mannoses and mannosylphosphates, whereas the formation of mammalian glycans generally involves the removal of mannose followed by the addition of N-acetylglucosamine, galactose, fucose, and sialic acid (*Figure. 8.1*) (Dean, 1999; Gemmill and Trimble, 1999; Kukuruzinska and Lennon, 1998; Moremen *et al.*, 1994).

Thus, even though yeasts are able to attach N-linked glycans to proteins, these high mannose glycans are of nonhuman nature and can negatively impact protein half-life due to their high affinity to the high mannose receptors present on macrophages and endothelial cells (Mistry *et al.*, 1996). In addition, yeast N-glycans can contain beta-linked mannoses (Davidson *et al.*, 2004; Vinogradov *et al.*, 2000) as well as mannosylphosphates which may increase the risk of an immune response.

Other eukaryotic expression systems, such as plant and insect cells are equally unable to provide mammalian glycosylation (Bardor *et al.*, 2003; Gomord and Faye, 2004; Kim *et al.*, 2005). For example, nonhuman sugars such as core α-1-3 fucose, present on proteins expressed in plants and insect cells, and xylose, present in plants, constitute important IgE epitopes. Clinical use of glycoproteins containing such sugars may be accompanied by severe adverse reactions, including life-threatening anaphylaxis, especially in allergic patient populations (Bardor *et al.*, 2003; Gomord and Faye, 2004). In addition, approximately 1% of circulating IgGs in normal human serum is composed of anti-α-1,3 galactose antibodies, a common modification to N-linked glycans produced by nonprimate mammals (Galili *et al.*, 1984).

For these reasons mammalian cells, in particular Chinese hamster ovary (CHO) cells, have emerged as the preferred expression system for the

Figure 8.1

Major N-glycosylation pathways in humans and yeast. Early N-glycosylation steps in the ER are conserved. Upon entry into the Golgi maturation of N-glycans diverges significantly between mammalian and fungal cells. Mns: α 1,2- mannosidase; MnsII: mannosidase II; GnT1: β-1,2-N-acetylglucosaminyltransferase I; GnTII: β-1,2-N-acetylglucosaminyltransferase II; GalT: β-1,4-galactosyltransferase; SiaT: α-2,6-sialyltransferase; MnT:

○, β-1,N-GlcNAc; ●, β-1,4-GlcNAc; ◐, β-1,2-GlcNAc; ▥, β-1,4-Man; ▨, α-1,6-Man; ▧, α-1,2-Man; ▤, α-1,3-Man; ◇, β-1,4-Gal; ✦, α-2,6-Sia production of complex human glycoproteins. However, it is commonly recognized that glycan structures produced from CHO cell lines differ from those produced in human cells and sometimes have to be modified to meet therapeutic efficacy (Grabenhorst *et al.*, 1999; Grabowski *et al.*, 1995). On the other hand, the presence of specific glycoforms can confer significant

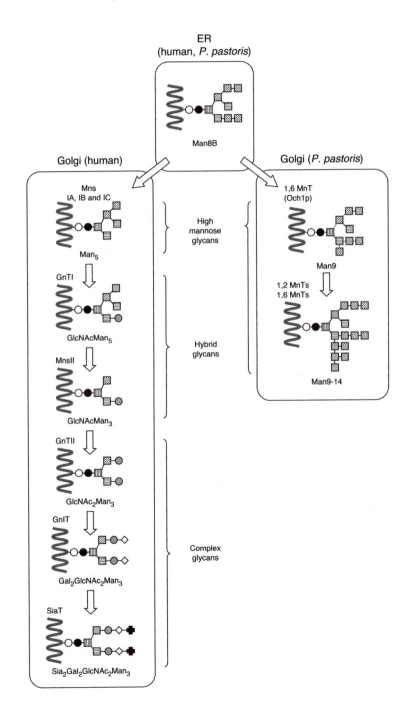

therapeutic advantages, and thus expression systems that allow for the control of glycosylation are desirable.

8.2 Yeast as host for recombinant protein expression

The recombinant protein production process can be divided into three stages that contribute comparably to the final cost: (i) protein production (i.e. fermentation), (ii) protein purification and (iii) the filling and packaging of a formulated final product. Due to the fixed cost nature of purification and formulation, most efforts to reduce cost have focused on the fermentation process. One approach to reduce the fermentation cost is the evaluation of various expression systems.

Yeast has proven to be a robust, scalable protein expression platform for many industrial enzymes and some nonglycosylated therapeutic proteins. Almost one sixth of the currently approved therapeutic proteins are made in yeast, including recominant GM-CSF, the hepatitis B, and several other vaccines (McAleer et al., 1984). More than half of the world's supply of insulin is made in yeast by Novo Nordisk (Novolog, Novo Nordisk, Bagsvaerd, Denmark). In addition, yeast based systems could offer a significant reduction in the drug development timeline due to their short doubling times, the ease of genetic manipulations and the wealth of information available on process development, cell biology and molecular genetics.

Fungal-type N-linked glycosylation inherent to yeast has, however, precluded their use for the production of therapeutic glycoproteins. To overcome this shortcoming, several research groups have investigated the possibility of altering endogenous N-glycosylation reactions and genetically engineering human N-glycosylation pathways into yeast and other fungi.

8.3 N-linked glycosylation overview: Fungal versus mammalian

Molecules containing carbohydrates (glycans) covalently linked to a noncarbohydrate moiety are classified as glycoconjugates, either glycoproteins or glycolipids. In yeasts and other eukaryotes, glycoproteins typically contain N- or O-linked glycans (Herscovics and Orlean, 1993; Strahl-Bolsinger et al., 1999). O-linked glycosylation is the attachment of carbohydrates to the hydroxyl-groups of serine or threonine residues of proteins. In contrast to N-glycosylation, not much is known about O-glycosylation and its biological role (reviewed in: Gentzsch and Tanner, 1997; Lehle and Bause, 1984; Strahl-Bolsinger et al., 1999).

Given the relevance of N-linked glycosylation with regard to the function of glycoproteins, this chapter focuses on efforts to modify N-glycosylation in yeast and other fungal protein expression hosts, although similar efforts are under way to modify and humanize O-glycosylation in yeast.

The early steps of N-linked glycosylation, residing in the endoplasmic reticulum (ER) are conserved between mammals and lower eukaryotes, such as yeast. In both mammals and yeast, the dolichol-linked core oligosaccharide ($Glc_3Man_9GlcNAc_2$) is transferred onto accessible Asn residues within the Asn-X-Ser/Thr (where X is any amino acid other than proline) motif by the multi-subunit enzyme complex oligosaccharyltransferase (OST). Following the transfer, three glucose residues and one terminal α-1,2 mannose residue are trimmed by glucosidase I, II and an ER-residing α-1,2 mannosidase, respectively. Glycoproteins with $Man_8GlcNAc_2$ N-linked oligosaccharides are then transported to the Golgi apparatus (Burda *et al.*, 1999). Some yeasts, such as *Schizosaccharomyces pombe* and *K. lactis* appear to lack ER-specific α-1,2 mannosidase activity, resulting in the transport of $Man_9GlcNAc_2$-containing glycoproteins to the Golgi (our observation and Gemmill and Trimble, 1999). In the Golgi N-glycan processing diverges significantly between yeast and mammals (*Figure 8.1*).

In humans and other mammals, one of the first early Golgi residing N-glycan processing steps involves the trimming of $Man_8GlcNAc_2$ to $Man_5GlcNAc_2$ by α-1,2-mannosidase(s). The resulting $Man_5GlcNAc_2$ N-linked oligosaccharide serves as the substrate for N-acetylglucosaminyl transferase I (GnTI), which transfers a single N-acetylglucosamine (GlcNAc) sugar onto the terminal 1,3-mannose of the tri-mannose core. The resulting N-linked $GlcNAcMan_5GlcNAc_2$ is recognized by mannosidase II which removes the two remaining α-1,3 and α-1,6 terminal mannose sugars to produce $GlcNAcMan_3GlcNAc_2$. This is the substrate for N-acetyl-glucosaminyl transferase II (GnTII) which adds one GlcNAc sugar to the terminal α-1,6-mannose arm of the tri-mannose core to yield $GlcNAc_2Man_3GlcNAc_2$ (see *Figure 8.1*) (Herscovics, 1999, 2001; Moremen and Robbins, 1991; Tabas and Kornfeld, 1979; Tulsiani *et al.*, 1982).

Terminal GlcNAc residues typically serve as substrate for galactose transfer, followed by the addition of terminal sialic acid, in particular N-acetyl-neuraminic acid (NANA) residues. A number of GlcNAc transferases capable of adding additional antennae exist (e.g. GnTIV and GnTV) (Brockhausen *et al.*, 1988; Schachter, 1991). Galactose and NANA transfer, resulting in tri- and tetra-antennary N-linked glycans, in turn can extend these antennae. Furthermore, a host of additional glycosyltransferases, including fucosyl-transferases, GalNAc transferases and GlcNAc-phosphotransferases, are known to exist.

It is important to note that these enzymes may compete with each other for their respective substrates. Thus, many of the reactions catalyzed by glycosydases and glycosyltransferases are not driven to completion and glycoproteins produced by mammalian cells as well as those purified from human serum typically display a significant number of intermediates ranging from high mannose to complex terminally sialylated glycans (Kornfeld and Kornfeld, 1985; Schachter, 2000).

In yeast a set of mannosyltransferases and mannosylphosphate trans-ferases add mannose and mannosylphosphate residues. In *S. cerevisiae*, α-1,2, α-1,3 and α-1,6 mannosyltransferases as well as mannosylphosphate transferases have been described and N-linked oligosaccharides display mannosylated and hyper-mannosylated N-glycan structures of up to 100 mannose residues (Ziegler *et al.*, 1994). In *K. lactis*, *H. polymorpha* and *P. pastoris*, a similar set of mannosyltransferases exists, resulting in mostly high mannose structures that are typically smaller than those produced by *S. cerevisiae*. In addition to the α-mannosyltransferases we have recently identified a group of β-mannosyltransferases in *P. pastoris*. The presence of β-linked mannoses on yeast produced therapeutic glycoproteins could trigger an immune response in humans and we have eliminated the responsible genes from the genome of the yeast *P. pastoris* (Davidson *et al.*, 2004).

8.4 A brief history of efforts to humanize N-linked glycosylation in fungal systems

As described above, in spite of nonhuman post-translational modifications, yeast and filamentous fungi offer attractive attributes as protein expression platforms. The most straightforward approach towards achieving the goal of human-like glycosylation in fungal systems is to genetically engineer the human glycosylation pathway into the host organism itself. As discussed above N-linked glycosylation is an extremely complex biosynthetic process. A multitude of interdependent enzymatic reactions requiring different substrates are lined up along the secretory pathway of a eukaryotic cell. Thus, replicating a human glycosylation pathway in a nonhuman host that contains its own endogenous set of glycosylation reactions, is a daunting task at best.

By the early 1990s, a significant body of literature existed describing important aspects of fungal and human glycosylation (Burda and Aebi, 1999; Kornfeld and Kornfeld, 1985; Ziegler *et al.*, 1994). Over the past few years various groups have explored the possibility of humanizing N-glycosylation pathways in fungal hosts to produce human-like glycoproteins. One of the first attempts to offer a fungal expression host as an alternative to mammalian cell lines was initiated after it was discovered that the filamentous fungus *Trichoderma reesei* was able to secrete glycoproteins containing intermediates of human glycosylation structures. Even more, *in vitro* processing experiments on these intermediates employing mammalian glycosyltransferases showed some promise (Maras and Contreras, 1994). A follow-up effort, published in 1997, consisted of employing a fungal precursor protein as a substrate for one or several chemoenzymatic glycosylation steps, resulting in hybrid glycoproteins, albeit at low yield (Maras *et al.*, 1997). One of the drawbacks of this approach was the increased cost and complexity associated with the need for several *in vitro* glycosylation reactions. However, the fact that *in vitro* modification proved feasible encouraged research towards high titer production of human glycoproteins in yeast and filamentous fungi.

In order to achieve uniform human-like complex glycans, each reaction and in particular the early steps have to be obtained at high yields, since the biosynthetic pathway of N-glycosylation is a series of interdependent reactions, where each step creates the substrate for the following enzymatic reaction (*Figure 8.1*). Of the early steps, the first critical step in the generation of complex glycans is the trimming of $Man_8GlcNAc_2$ to $Man_5GlcNAc_2$ in the Golgi apparatus by α-1,2 mannosidase(s) (*Figure 8.1*).

In 1995, Harry Schachter's group at the Sick Children's Hospital of the University of Toronto reasoned that, based on the activity of an endogenous α-1,2 mannosidase in the filamentous fungus *Aspergillus nidulans*, human-like N-glycans could be obtained by co-expression with a heterologous GnTI. The α-1,2 mannosidase was expected to trim the $Man_8GlcNAc_2$ to $Man_5GlcNAc_2$, while GnTI would add a *N*-acetylglucosamine (GlcNAc) sugar onto the terminal 1,3-mannose of the tri-mannose core, producing a human-like hybrid glycan (GlcNAcMan$_5$GlcNAc$_2$). However, while high GnTI transcription levels were measured and cell-free extracts contained active enzyme, no human-like N-linked glycans were detected (Kalsner *et al.*, 1995).

There exist a number of possible explanations why no transfer of *N*-acetylglucosamine was achieved, some of which are: (i) GnTI may not have been targeted to the correct locality within the secretory pathway; (ii) no Man$_5$GlcNAc$_2$ acceptor substrate may have been present at the site of GnTI localization; (iii) the micro-environment within the secretory pathway is not conducive to GnTI activity (e.g. incorrect pH or temperature); and (iv) no, or insufficient, amounts of UDP-GlcNAc at the site of GnTI localization were present, or any combination of the above could have been responsible.

One of the first efforts to achieve *in vivo* N-glycan processing in yeast was an attempt to localize mannosidases in the secretory pathway of *P. pastoris*. In 1998, Martinet and co-workers demonstrated that an α-1,2 mannosidase from *T. reesei*, containing a putative N-terminal transmembrane domain, was not efficiently retained when expressed in *P. pastoris* (Martinet *et al.*, 1998). Martinet *et al.* then fused the catalytic domain of the same enzyme to the secretory pathway targeting sequence derived from the *S. cerevisiae* Golgi residing protein Mns1p. This fusion prevented secretion but unfortunately failed to result in mannose trimming (Martinet *et al.*, 1998). The same group followed another approach in 1998 to create Man$_5$GlcNAc$_2$ N-linked oligosaccharides in the yeast *Pichia pastoris*. An attempt to express an active α-1,2 mannosidase resulted in even higher molecular weight glycans; most likely because *Pichia pastoris* appeared to compensate for mannose trimming by up-regulating endogenous mannosyltransferases (Martinet *et al.*, 1998). These results highlight the difficulties experienced in the empirical approaches taken to re-engineer N-glycosylation in fungal hosts.

8.5 Sequential targeting of glycosylation enzymes is a key factor

In yeast and humans host-specific glycosyltransferases and glycosidases line the luminal surface of the ER and Golgi apparatus. As a glycoprotein moves along the secretory pathway these enzymes provide the catalytic surface along which the sequential processing and maturation of *N*-glycans takes place. As described above, several processing steps rely on the previous upstream reactions. It is therefore important that N-glycosylation enzymes are arranged in a predetermined sequence along the secretory pathway.

The secretory pathway can roughly be divided into compartments through which a protein destined for secretion will proceed (from the ER, to the cis and medial Golgi, and finally across the trans-Golgi network into the medium). The exact processes or mechanisms of protein transport through the secretory pathway, while each compartment maintains a specific set of resident enzymes, remain an extensive research subject (Gleeson, 1998; Harter and Wieland, 1996; van Vliet *et al.*, 2003).

Among the better-understood mechanisms responsible for the localization of proteins to various compartments are retrieval and retention (reviewed in: Teasdale and Jackson, 1996; van Vliet *et al.*, 2003). In the case of retrieval protein–protein interactions are the underlying retention mechanism allowing for localization. For example, some ER-residing proteins display a carboxy-terminal tetrapeptide with the consensus sequence KDEL (or HDEL in yeast) that has been shown to be required for efficient ER localization (Munro and Pelham, 1987; Pelham *et al.*, 1988). The receptor protein, Erd2p binds to the K/HDEL tetrapeptide in the Golgi. Upon retrograde transport back into the lower pH ER, the protein is released from the protein–receptor complex (Dean and Pelham, 1990; Lewis *et al.*, 1990; Semenza *et al.*, 1990).

A number of ER- and Golgi-residing enzymes (for example α-1,2 mannosidase, mannosidase II and GnT I) display a common modular structure and are classified as type II membrane proteins (*Figure 8.2*). These proteins have a short cytoplasmic tail at the amino terminus, a hydrophobic transmembrane domain and a luminal stem followed by a carboxy-terminal catalytic domain (Paulson and Colley, 1989) Even though deletion studies as well as fusions to nonGolgi-residing proteins have established that the targeting information of many type II membrane proteins resides within the amino-terminus, there exists some ambiguity as to whether this targeting information is conserved across different species. For example, human 2,6 sialyl transferase (2,6 ST) and galactosyl transferase (GalT) which are trans-Golgi and late-Golgi residing enzymes, respectively, have been shown to be localized to the ER of *S. cerevisiae* when expressed recombinantly (Krezdorn *et al.*, 1994). However, fusion of the N-terminal trans-membrane domain of rat 2,6 ST to yeast invertase, a secreted glycoprotein, localized this enzyme to the Golgi of *S. cerevisiae* (Schwientek and Ernst, 1994). While these and other studies have established a correlation between certain domains and targeting events, there is currently no tool available that allows one to predict the behavior of a given targeting peptide across different hosts.

Notwithstanding the underlying complexity, researchers have attempted to exploit targeting mechanisms to establish human-like N-glycosylation *in vivo* in yeast. As mentioned above Martinet and co-workers found that α-1,2

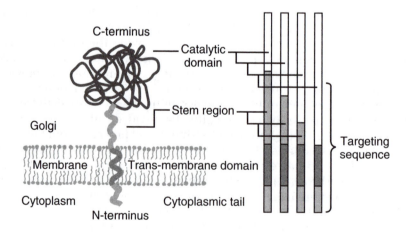

Figure 8.2

Type II trans-membrane proteins. Most secretory pathway residing glycosidases and glycosyltransferases are anchored to the membrane through a type II transmembrane anchor.

mannosidase from *T. reesei* fused to a putative N-terminal transmembrane domain did not allow sufficient retention of the enzyme in the secretory pathway of *P. pastoris*. However, when the same enzyme was fused to the leader sequence of *S. cerevisae* Mns1p secretion was prevented, although no mannose trimming was demonstrated (Martinet *et al.*, 1998).

The first successful attempt to generate human-like *N*-glycans *in vivo* in yeast was reported by Chiba and co-workers in collaboration with researchers at Kirin (Yokohama, Japan). This group launched a systematic approach and eliminated *OCH1*, *MNN11* and *MNN4* (all of these genes encode for Golgi residing glycosyltransferases) in the yeast *S. cerevisiae*. They then employed the leader sequence of *S. cerevisae* Och1p, a Golgi residing α-1,6 mannosyltransferase, to localize the α-1,2 mannosidase from *A. saitoi* to the Golgi of *S. cerevisae*. Although no trimming of $Man_8GlcNAc_2$ to $Man_5GlcNAc_2$ was observed, they were able to demonstrate Golgi, as well as cytosol targeting of the α-1,2 mannosidase. It was only when an HDEL retrieval mechanism was used to localize the catalytic domain of the α-1,2 mannosidase to the ER that some mannosidase activity could be detected on an intracellular reporter protein. However, even though a 2μ multicopy plasmid with a strong constitutive promoter (GAPDH) driving α-1,2 mannosidase was employed, this reaction yielded less than 30% $Man_5GlcNAc_2$ (Chiba *et al.*, 1998).

The processing of $Man_8GlcNAc_2$ to $Man_5GlcNAc_2$ is only one early step in the maturation of human-like *N*-glycans. Given the large number of subsequent steps needed to obtain a fully matured complex human-like N-linked oligosaccharide this low yield was of concern and highlighted the fact that despite all of its promises the *in vivo* humanization of N-linked glycosylation in fungal hosts was an ambitious undertaking. Merely deleting endogenous genes involved in N-glycosylation and expressing enzymes involved in mammalian glycosylation appeared to offer little if any chances of being successful.

8.6 Replication of human-like glycosylation in the methylotrophic yeast *Pichia pastoris*

An essential step in recreating human glycosylation pathways in lower eukaryotes involves the ability to recreate the sequential nature of complex *N*-glycan processing. By 2000, our group launched a research program with three objectives. It was our intention to (i) eliminate the endogenous nonhuman yeast glycosylation enzymes responsible for the addition of the outer chain mannoses and other undesired sugar residues; (ii) develop methods to identify yeast strains harboring targeted heterologous enzymes that are active at the desired site of localization within the secretory pathway; and (iii) develop efficient screens to identify those strains which have acquired the ability to secrete and process recombinant glycoproteins in a human-like fashion at high yield.

8.7 A library of α-1,2 mannosidases

The first maturation step of mammalian *N*-glycans is the trimming of $Man_8GlcNAc_2$ to $Man_5GlcNAc_2$ by α-1,2 mannosidases. To demonstrate successful mannose trimming we created a Δ*och1* strain secreting a hexahis-tidine-tagged fragment of human plasminogen as a reporter protein. *OCH1* is an initiating α-1,6 mannosyltransferase and Δ*och1* strains lack the ability to synthesize the so-called outer chain (*Figure 8.1*). To ensure the correct targeting of the active heterologous mannosidase enzyme, we designed a genetic library of N-terminal domains derived from known type II membrane proteins, naturally localized to the ER and Golgi of *S. cerevisiae* or *P. pastoris* (Choi *et al.*, 2003).

This targeting sequence library includes N-termini from type II membrane proteins such as Gls1p, Mns1p, Sec12p, Mnn9p, Van1p, Anp1p, Hoc1p, Mnn10p and Mnn11p from *S. cerevisiae* as well as Och1p and Sec12p from *P. pastoris*. A second genetic library harbored catalytic domains of several α-1,2 mannosidases (lacking the N-terminal fragments including the trans-membrane domain) from *Homo sapiens*, *Mus musculus*, *Aspergillus nidulans*, *C. elegans*, *Drosophila melanogaster*, and *Penicillium citrinium*. These libraries were designed in a way that simple ligation of the targeting and catalytic domains resulted in a combinatorial library encoding for chimeric fusions between all the yeast derived targeting sequences and all the catalytic domains. This α-1,2 mannosidase library consisted of 608 chimeric fusion constructs. The library was transformed into the Δ*och1* host strain secreting the reporter protein. Secreted protein was affinity purified, *N*-glycans released by digestion with protein *N*-glycanase and analyzed by matrix assisted laser desorption/ionization – time of flight (MALDI-TOF) mass spectroscopy. Only two strains harbored chimeric α-1,2 mannosi-dases, efficiently trimming (>90%) *in vivo* $Man_8GlcNAc_2$ to $Man_5GlcNAc_2$.

The fact that only such a small fraction of constructs achieved the desired specific activity further highlights that proper intracellular targeting and desired enzyme activity may be heavily influenced by the specific interaction between targeting sequences and catalytic domains.

8.8 Transfer of *N*-acetylglucosamine

In mammalian cells the terminal α-1,3 mannose of an N-linked $Man_5GlcNAc_2$ glycan can receive an *N*-acetylglucosamine to generate $GlcNAcMan_5GlcNAc_2$ (*Figure 8.1*). The early Golgi residing enzyme *N*-acetyl glucosamine transferase I (GnTI) catalyzes the transfer reaction of GlcNAc from the sugar donor UDP-GlcNAc onto N-linked $Man_5GlcNAc_2$. As mentioned above, Schachter's group attempted to introduce a GnTI from rabbit into the filamentous fungus *A. nidulans*. This group was not able to detect the desired product $GlcNAcMan_5GlcNAc_2$ (Kalsner *et al.*, 1995).

Given the previously unsuccessful attempts to produce $GlcNAcMan_5GlcNAc$ in fungal systems, we reasoned that in order to achieve successful GlcNAc transfer, three requirements had to be considered: (i) the necessary substrate $Man_5GlcNAc_2$ had to be present; (ii) an active GnTI enzyme, targeted to the correct locality within the secretory pathway, was necessary; and (iii) sufficient amounts of the nucleotide sugar UDP-GlcNAc needed to be made available within the Golgi.

We assumed that a pool of UDP-GlcNAc exists within the cytosol, since the assembly of dolichol-linked glycan precursors, which takes place on the cytosolic side of the ER, requires this nucleotide sugar.

Having obtained selected strains displaying highly uniform Man$_5$GlcNAc$_2$ N-glycans we proceeded to generate strains expressing the Golgi UDP-GlcNAc transporter derived from the yeast *Kluyveromyces lactis* in order to create a pool of UDP-GlcNAc within the Golgi. Next, we screened chimeric GlcNAcT1 fusions, created in a fashion similar to the one described above for the α-1,2 mannosidase library step, and obtained strains capable of glycosylating the secreted reporter protein in an essentially uniform manner with GlcNAcMan$_5$GlcNAc$_2$ N-glycans (Choi *et al.*, 2003). We also established that the presence of a UDP-GlcNAc transporter is necessary to achieve complete conversion of Man$_5$GlcNAc$_2$ to GlcNAcMan$_5$GlcNAc$_2$.

8.9 Two independent approaches towards complex *N*-glycans: How to eliminate more mannoses

GlcNAcMan$_5$GlcNac$_2$ has two terminal mannoses, one an α-1,3 and the other an α-1,6 mannose. Mannosidase II, a 131-kDa Golgi-resident protein, recognizes GlcNAcMan$_5$GlcNAc$_2$ as a substrate and cleaves the α-1,3- and the α-1,6- mannose to yield GlcNAcMan$_3$GlcNAc$_2$ (*Figures 8.1 and 8.3*). This trimming reaction depends on a GlcNAc residue being present on the α-1,3 arm of the trimannose core. Thus, in order to achieve an efficient mannose cleavage event, mannosidase II activity has to be co-localized with or downstream of the GnTI catalyzed step in the secretory pathway.

Due to the uncertainties associated with the expression of the large recombinant mannosidase II enzyme, we opted to pursue two parallel approaches. In one approach we used our established library technology to attempt active mannosidase II expression and targeting, while in the other one we interfered with the early stages of lipid-linked oligosaccharide assembly.

In *S. cerevisiae* mutants had been identified transferring truncated oligosaccharides onto glycoproteins. One particular gene, *ALG3*, encodes for a mannosyltransferase responsible for the addition of the first α-1,3 mannose to the α-1,6-arm of the trimannose core (*Figure 8.3*). An Alg3p deficient yeast will not add the first α-1,3 mannose. In addition, the subsequent step, addition of the α-1,6 mannose to the α-1,6-arm of the trimannose core will be prevented. Thus, exactly those two mannose residues, which mannosidase II will trim in the process of mammalian N-glycan maturation are not present in Δ*alg3* mutant yeast (Aebi *et al.*, 1996; Sharma *et al.*, 2001). We identified, cloned and eliminated *ALG3* from the genome of *P. pastoris* (Davidson *et al.*, 2004). We reasoned that in the absence of the outer-chain initiating mannosyltransferase Och1p, (*Figures 8.1 and 8.3*) the resulting Δ*alg3* specific Man$_5$GlcNAc$_2$ N-glycan could be a substrate for α-1,2 mannosidase to yield paucimannose (Man$_3$GlcNAc$_2$) N-linked glycans (*Figure 8.3*). Earlier work by Schachter's group demonstrated that paucimannose can serve as an *in vitro* substrate for GnTI (Brockhausen *et al.*, 1988). We postulated that the same could be replicated *in vivo*. In summary, our strategy involved creating a Δ*alg3* Δ*och1* yeast harboring an active and properly localized combination of α-1,2 mannosidase with GnTI catalyzed GlcNAc transfer to circumvent the need for mannosidase II.

However it was not known (i) to what extent an *alg3*, *och1* double mutant *P. pastoris* strain would be viable; (ii) to what extent the *alg3* deletion is sufficiently tight to prevent subsequent addition of α-1,6-mannose by Alg12p; and (iii) to what extent this lipid-linked glycan will be recognized by oligosaccharyltransferase and thus impact *N*-glycan occupancy.

We have shown subsequently that a viable *alg3, och1* double mutant of *P. pastoris* can be obtained and that introduction of active α-1,2 mannosidase, GnTI, GnTII (the enzyme adding a GlcNAc to the 1,6 mannose of a GlcNacMan₃GlcNAc₂ arm) and a UDP-GlcNAc transporter allow for the generation of uniform complex glycoproteins with terminal GlcNAc (*Figure 8.3* and Bobrowicz *et al.*, 2004). The parallel approach of identifying strains

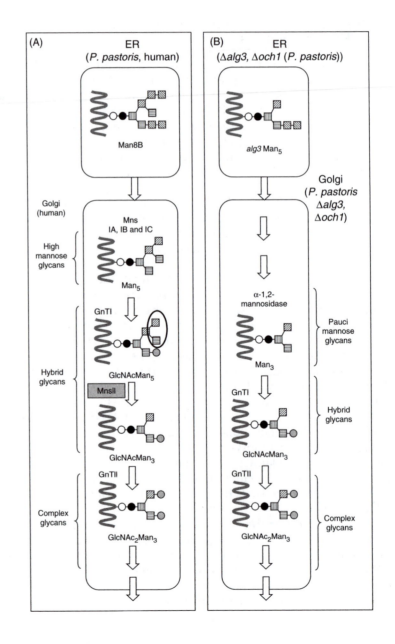

with active and properly localized mannosidase II and N-acetylglucosamine transferase II was also successful and yielded *P. pastoris* strains with the ability to secrete uniform complex $GlcNAc_2Man_3GlcNAc_2$ glycans (Hamilton *et al.*, 2003).

8.10 Some metabolic engineering: Transfer of galactose

Strains secreting recombinant proteins with terminal GlcNAc glycan provide a substrate for transfer of galactose, the next step in mammalian glycosylation. The requirements for transfer of galactose are quite similar to those for successful GlcNAc transfer: a protein with *N*-glycans with terminal GlcNAc as a substrate, a Golgi pool of UDP-galactose and a properly targeted and active β-1,4 galactose transferase (β-1,4 GalT).

A very elegant approach published by Schwientek *et al.* suggested that galactose transfer could be achieved *in vivo* in the yeast *S. cerevisiae*. Schwientek and co-workers created a yeast mutant deficient in *alg1*, one of the genes involved in lipid-linked oligosaccharide assembly, which at nonpermissive temperatures displays short N-linked $GlcNAc_2$ stumps. This *alg1* mutant yeast was shown to provide an artificial substrate for GalT, albeit at low efficiency (Schwientek *et al.*, 1996).

Recently, the group headed by Contreras demonstrated some galactose transfer activity in *P. pastoris*. This group introduced a β-1,4-GalT targeted to the Golgi apparatus of a *P. pastoris* strains producing $GlcNAcMan_5GlcNAc_2$ N-linked glycans. As was in the case for Schwientek *et al.*, the transfer of galactose was only partial (Vervecken *et al.*, 2004). We recently demonstrated nearly quantitative transfer of galactose on complex glycoproteins in the yeast *P. pastoris*. Two approaches were pursued. Both of which are based on host cells that produce bi-antennary structures that terminate in GlcNAc (Bobrowicz *et al.*, 2004; Hamilton *et al.*, 2003). In one approach we screened a library of GalT/targeting sequence fusions. Even though we identified constructs achieving significantly higher transfer than previously published, we did not find strains displaying complete transfer of galactose with any of the tested constructs.

Figure 8.3 (opposite)

Engineering of a Δ*alg3*, Δ*och1 P. pastoris* mutant strain to produce complex glycans. (A) Modification of *N*-glycans in mammalian cells. The two mannose residues trimmed by mannosidase II (and not present in a Δ*alg3*, Δ*och1 P. pastoris* mutant strain) are circled. (B) N-glycosylation pathway in an engineered Δ*alg3*, Δ*och1 P. pastoris* mutant strain containing heterologous mannosidase and glycosyltransferases. This strain is capable of producing complex human *N*-glycans through deletion of *PpOCH1* and expression of Mns I and GnT I as described previously (Choi *et al.*, 2003), followed by deletion of *PpALG3* and expression of GnT II. Mns: α-1,2-mannosidase I; Mns II: α-mannosidase II; GnT I: β-1,2-*N*-acetylglucosaminyltransferase I; GnT II: β-1,2-*N*-acetylglucosaminyltransferase II; MnT: mannosyltransferase.

○, β-1,*N*-GlcNAc; ●, β-1,4-GlcNAc; ◐, β-1,2-GlcNAc; ▥, β-1,4-Man; ▨, α-1,6-Man; ▧, α-1,2-Man; ▤, α-1,3-Man

The presence of a UDP-galactose transporter in the yeast *S. cervevisiae* has been described previously (Roy *et al.*, 1998). We postulated that the intra-cellular and, in particular, the Golgi pool of UDP-galactose may be insufficient in *P. pastoris* and thus limit galactose transfer. To overcome this limitation we cloned a gene from the yeast *Schizosaccharomyces pombe* with strong homology to known UDP-galactose-4-epimerases. *Pichia pastoris* is a yeast that is not able to assimilate galactose. Galactose-utilizing organisms (including microorganisms and mammals) employ UDP-galactose-4-epimerase for the reversible conversion of UDP-glucose and UDP-galactose. Upon cytosolic co-expression of this enzyme we found a significant improvement in galactose transfer but were still not satisfied with the degree of galactosylation achieved. Only after co-expressing an UDP-galactose transporter from *Drososphila melanogaster*, the *S. pombe* UDP-galactose-4-epimerase and GalT we accomplished galactose transfer to near homogeneity (*Figure 8.4* and Davidson *et al.*, 2004).

To alleviate the risk that a heterologous sugar nucleotide transporter with multiple transmembrane domains may not function, a UDP-galactose-4-epimerase GalT catalytic domain of GalT and a yeast localization leader fusion protein was created. This Golgi targeted fusion protein accomplished within the Golgi the conversion of UDP-glucose to UDP galactose as well as the transfer of galactose in the absence of a transporter (Bobrowicz *et al.*, 2004).

8.11 More metabolic engineering: Sialic acid transfer. The final step

In the process of mammalian N-glycosylation terminal galactose residues can be capped with sialic acid. In mammalian cells CMP-sialic acid, specifically CMP-*N*-acetylneuraminic acid, (CMP-NANA) is transferred. However, yeast lack the biosynthetic pathway to produce CMP-NANA as well as the transporter required to shuttle this nucleotide sugar into the Golgi as well as a properly targeted and active sialyltransferase (ST). Previous work described the production of secreted and active STs in yeast (Malissard *et al.*, 1999, 2000).

In a first step we created *P. pastoris* strains harboring genes of biosynthetic CMP-NANA pathways. We demonstrated CMP-NANA production *in vivo* in the cytosol of *P. pastoris* (unpublished). Strains harboring the functional CMP-NANA biosynthetic pathway and an appropriate CMP-NANA Golgi transporter allowed the use of a library-based screen to find the most efficient targeting domain / catalytic ST domain combination resulting in the secretion of glycoproteins with over 90% terminal sialylation.

In summary, generation of these strains capable of fully human-like N-glycosylation required the deletion of four genes to eliminate fungal-specific glycosylation and the introduction of fourteen heterologous genes (Hamilton *et al.*, 2006).

8.12 Glyco-engineered yeast as a host for production of therapeutic glycoproteins

As mentioned above protein production in yeast can offer several advantages: protein production processes are usually robust and scalable. In

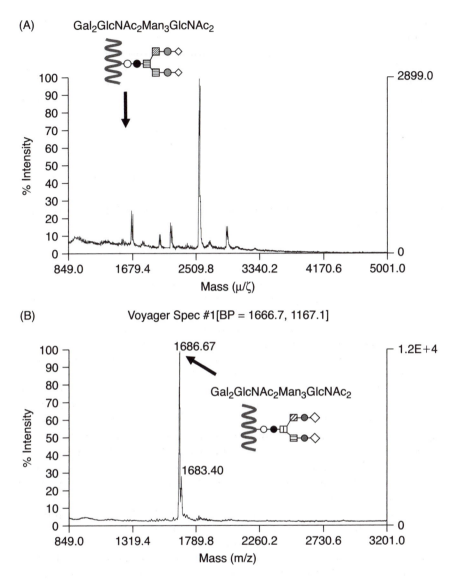

Figure 8.4

Example for *N*-glycan homogeneity of proteins secreted by glyco-engineered yeast. (A) MALDI-TOF of glycans released from commercial mammalian cell produced EPO. EPO was treated with neuroaminidase to remove terminal sialic acid. Glycans were released and analyzed. N-linked glycans are heterogeneous and consist of a mixture of glycans with multiple antennae and differing degrees of complexity. The arrow indicates the mass corresponding to the biantennary *N*-glycan $Gal_2GlcNAc_2Man_3GlcNAc_2$. (B) Rat EPO secreted by a glycoengineered *P. pastoris* strain. The arrow indicates the mass corresponding to the biantennary *N*-glycan $Gal_2GlcNAc_2Man_3GlcNAc_2$.

addition short cycle times and achievable productivity are very attractive attributes of yeast based fermentation processes. In particular the methylotrophic yeast *P. pastoris* has received increasing attention. Several *P. pastoris* produced therapeutic proteins are in varying stages of clinical trials (Gerngross, 2004). As described above our group has over the past years successfully glycoengineered *P. pastoris* to perform human-like glycosylation at a high degree of uniformity (Bobrowicz *et al.*, 2004; Choi *et al.*, 2003; Hamilton *et al.*, 2003, 2006). N-glycosylation plays a critical role in function and efficacy of glycoproteins. In the following section we will briefly discuss examples for the important role N-glycosylation plays with regard to pharmacokinetics, tissue targeting and potency of therapeutic glycoproteins.

8.13 N-linked glycans and pharmacokinetics of therapeutic glycoproteins

The composition and structure of N-linked glycans contribute to the molecular size, the hydrodynamic volume, and net charge of glycoproteins in circulation. All of these properties can influence PK, which in turn plays a major role in determining the circulating half-life and efficacy of many therapeutic proteins.

The content of sialic acid is a key determining factor in the net negative charge of glycoproteins. An increase in sialic acid content has been shown to contribute to an improvement of the PK of therapeutic glycoproteins such as erythropoietin (EPO) (MacDougall, 2002).

In contrast, the presence of some *N*-glycans may reduce plasma half-life by triggering lectin-mediated clearance mechanisms. For example, the asialoglycoprotein receptor (ASGPR) found in the liver will efficiently clear glycoproteins displaying *N*-glycans lacking terminal sialic acid (Fukuda *et al.*, 1989; Stockert, 1995). A second clearance mechanism, the reticulo-endothelial system, employs several high affinity receptors to bind and remove glycoproteins with both terminal mannose and GlcNAc (Maynard and Baenziger, 1981). Even though deglycosylated EPO shows a three-fold increased potency over its glycosylated counterpart, *in vitro* this molecule is so short-lived in circulation that it has no *in vivo* efficacy (Dordal *et al.*, 1985; Higuchi *et al.*, 1992).

Recently the *in vivo* potency of recombinant EPO was increased through protein engineering. Two additional N-linked sites were added (Elliott *et al.*, 2004) and although this engineered EPO has weaker *in vitro* receptor binding it displays extended serum half-life, resulting in increased efficacy (Egrie and Browne, 2001).

8.14 *N*-glycans and their role in tissue targeting of glycoproteins

Carbohydrate binding proteins, or lectins, are differentially expressed on different cell types and thus different tissues have varying affinities for specific glycoforms. Depending on the *N*-glycans displayed therapeutic glycoproteins can be targeted to specific cells and tissues.

Glucocerebrosidase (GBA) is a therapeutic glycoprotein approved for the treatment of Gaucher's disease. GBA relies on a glycan-dependent targeting mechanism. Recombinant GBA is produced by mammalian cells and undergoes significant *in vitro* downstream processing. The terminal sialic acid residues are removed by neuraminidase digest, followed by digestion with galactosidase to remove the terminal galactose residues. Finally, yet another *in vitro* digest with hexosaminidase removes the terminal GlcNAcs. The result is a GBA molecule with terminal paucimannose $Man_3GlcNAc$ *N*-glycans.

Liver residing macrophages display mannose-binding lectins (MBLs) to which *in vitro* glycoengineered GBA with terminal mannose *N*-glycans is targeted. In the liver the recombinant enzyme then metabolizes accumulated glucocerebroside (Friedman *et al.*, 1999).

8.15 *N*-glycans can modulate the biological activity of therapeutic glycoproteins

N-glycans can play a role in affecting the biological activity of glycoproteins. For example, N-glycosylation has been shown to impact the affinity of glycoproteins to their respective receptors. In the case of immnunoglobulins (Igs), each heavy chain contains one N-linked glycan in the constant region. These *N*-glycans are buried in a horseshoe-shaped cavity between the two heavy chains. Even though these glycans seem not to be exposed, it has been shown that glycosylation is essential for efficient interaction with Fc receptors (FcR). Certain glycoforms enhance FcR-mediated effector functions, including antibody dependent cell cytotoxicity (ADCC) and complement dependent cytotoxicity (CDC). In contrast, aglycosylated Igs display significantly reduced effector functions. In mammalian cells the use of a set of N-glycosylation inhibitors has allowed for the production of IgGs with varying *N*-glycan composition. The different preparations displayed a modulation of ADCC (Rothman *et al.*, 1989).

Our group recently demonstrated that glycoengineered yeast cell-lines can produce IgGs with human *N*-glycan structures at a high degree of uniformity (see below). Depending on the *N*-glycan attached, antibody mediated effector functions can be optimized (Li *et al.*, 2006).

8.16 Control of N-glycosylation offers advantages

Glycoproteins produced in mammalian cell culture display a very heterogeneous N-linked glycan profile (*Figure 8.4*). In contrast, we have engineered the secretory pathway of *P. pastoris* to secrete glycoproteins with high glycan uniformity (for example: *Figure 8.4* and Bobrowicz *et al.*, 2004; Choi *et al.*, 2003; Hamilton *et al.*, 2003, 2006). As discussed above, N-glycosylation is a critical determinant of glycoprotein half-life, tissue targeting and potency of therapeutic glycoproteins. Therefore, the ability to produce a glycoprotein with a given predetermined *N*-glycan structure creates the opportunity to explore the role distinct *N*-glycans play with regard to the activity and function of the glycoprotein. Exploiting this structure–activity relationship in a systematic way can pave the way towards the creation of specific glycoforms with improved therapeutic properties.

Most importantly, once a strain has been identified it can be scaled up and manufacturing can be performed on the same platform, allowing for the large-scale production of specific glycoforms.

As mentioned above we recently employed a panel of glycoengineered *P. pastoris* strains to produce various glycoforms of the monoclonal antibody Rituxan anti CD20 (Li *et al.*, 2006). Rituxan is used for the treatment of nonHodgkin's B-cell lymphoma, a cancer of the immune system. Rituxan targets the CD20 receptor on B cells. We employed a library of glycoengineered antibodies sharing an identical amino acid sequence with commercial Rituxan. Yet, specific glycoforms of Rituxan displayed an approximately 100-fold higher affinity to FcγRIII. FcγRIII is an Fc receptor relevant for triggering ADCC. In addition we were able to show that specific glycoforms of this antibody display improved human B-cell depletion *in vitro* (Li *et al.*, 2006).

Glycoengineered yeast can also be used to produce specific glycoforms of therapeutic glycoproteins, which could be efficiently targeted to defined tissues. For example, as discussed above, GBA produced in mammalian cells has to undergo significant downstream *in vitro* glycan-modification to achieve the desired paucimannose *N*-glycan structures with terminal mannose. Thus, the production and secretion of paucimannose N-glycosylated GBA could offer an advantage over current methods.

8.17 Conclusions

We believe that glycoengineered yeast will allow the systematically probing for N-glycosylation dependent therapeutic effects. Employing a library of yeast with predetermined N-glycosylation will provide a greater understanding of structure-function relationship. This in turn should result in the development of improved and novel therapeutics.

To date, most of the effort directed at protein engineering has primarily focused on variations of the primary amino acid sequence. Identifying *N*-glycan structures resulting in improved efficacy of therapeutic glycoproteins in the past has been restricted to either *in vitro* modifications, the addition of selected N-glycosylation inhibitors (resulting in still heterogeneous pools of glycoforms) or cumbersome purification efforts to enrich for particular glycoforms.

We are currently expanding our library of glycoengineered yeast strains. We intend to create an array of strains, each capable of secreting a particular glycoform of proteins of interest. Once a glycovariant has been identified, scale-up and production at a commercial scale of hitherto unobtainable glycoforms of therapeutic proteins should become a reality.

References

Aebi M, Gassenhuber J, Domdey H and Te Heesen S (1996) Cloning and characterization of the ALG3 gene of *Saccharomyces cerevisiae*. *Glycobiology* 6: 439–444.

Bardor M, *et al.* (2003) Immunoreactivity in mammals of two typical plant glyco-epitopes, core alpha(1,3)-fucose and core xylose. *Glycobiology* 13: 427–434.

Bobrowicz P, Davidson RC, Li H *et al.* (2004) Engineering of an artificial glycosylation pathway blocked in core oligosaccharide assembly in the yeast *Pichia pastoris*: production of complex humanized glycoproteins with terminal galactose. *Glycobiology* 14: 757–766.

Brockhausen I, Narasimham S and Schachter H (1988) The biosynthesis of highly branched N-glycans: studies on the sequential pathway and functional role of

N-acetylglucosaminyltransferases I, II, III, IV, V and VI. *Biochimie* **70**: 1521–1533.

Burda P and Aebi M (1999) The dolichol pathway of N-linked glycosylation. *Biochim Biophys Acta Gen Subj* **1426**: 239–257.

Burda P, Jakob CA, Beinhauer J, Hegemann JH and Aebi M (1999) Ordered assembly of the asymmetrically branched lipid-linked oligosaccharide in the endoplasmic reticulum is ensured by the substrate specificity of the individual glycosyltransferases. *Glycobiology* **9**: 617–625.

Chiba Y, Suzuki M, Yoshida S, Yoshida A, Ikenaga H, Takeuchi M, Jigami Y and Ichishima K (1998) Production of human compatible high mannose-type (Man(5)GlcNAc(2)) sugar chains in *Saccharomyces cerevisiae*. *J Biol Chem* **273**: 26298–26304.

Choi BK, Bobrowicz P, Davidson RC *et al.* (2003) Use of combinatorial genetic libraries to humanize N-linked glycosylation in the yeast *Pichia pastoris*. *Proc Natl Acad Sci USA* **100**: 5022–5027.

Coloma MJ, Clift A, Wims L and Morrison SL (2000) The role of carbohydrate in the assembly and function of polymeric IgG. *Mol Immunol* **37**: 1081–1090.

Davidson RC, Nett JH, Renfer E *et al.* (2004) Functional analysis of the ALG3 gene encoding the Dol-P-Man: Man5GlcNAc2-PP-Dol mannosyltransferase enzyme of *P. pastoris*. *Glycobiology* **14**: 399–407.

Dean N. (1999) Asparagine-linked glycosylation in the yeast Golgi. *Biochim Biophys Acta Gen Subj* **1426**: 309–322.

Dean N and Pelham HRB (1990) Recycling of proteins from the Golgi compartment to the Er in yeast. *J Cell Biol* **111**: 369–377.

Dordal M, Wang F and Goldwasser E (1985) The role of carbohydrate in erythropoietin action. *Endocrinology* **116**: 2293–2299.

Egrie JC and Browne JK (2001) Development and characterization of novel erythropoiesis stimulating protein (NESP). *Nephrol Dialysis Transplant* **16(Suppl 3)**: 3–13.

Elliott S, Egrie J, Browne J, Lorenzini T, Busse L, Rogers N and Ponting I (2004) Control of rHuEPO biological activity: the role of carbohydrate. *Exp Hematol* **32**: 1146–1155.

Friedman B, Vaddi K, Preston C, Mahon E, Cataldo JR and McPherson JM (1999) A comparison of the pharmacological properties of carbohydrate remodeled recombinant and placental-derived beta-glucocerebrosidase: implications for clinical efficacy in treatment of Gaucher disease. *Blood* **93**: 2807–2816.

Fukuda M, Sasaki H, Lopez L and Fukuda M (1989) Survival of recombinant erythropoietin in the circulation: the role of carbohydrates. *Blood* **73**: 84–89.

Galili U, Rachmilewitz EA, Peleg A and Flechner I (1984) A unique natural human IgG antibody with anti-alpha-galactosyl specificity. *J Exp Med* **160**: 1519–1531.

Gemmill TR and Trimble RB (1999) Overview of N- and O-linked oligosaccharide structures found in various yeast species. *Biochim Biophys Acta* **1426**: 227–237.

Gentzsch M and Tanner W (1997) Protein-O-glycosylation in yeast: Protein-specific mannosyltransferases. *Glycobiology* **7**: 481–486.

Gerngross TU (2004) Advances in the production of human therapeutic proteins in yeasts and filamentous fungi. *Nat Biotechnol* **22**: 1409–1414.

Gleeson PA (1998) Targeting of proteins to the Golgi apparatus. *Histochem Cell Biol* **109**: 517–532.

Gomord V and Faye L (2004) Posttranslational modification of therapeutic proteins in plants. *Curr Opin Plant Biol* **7**: 171–181.

Grabenhorst E, Schlenke P, Pohl S, Nimtz M and Conradt HS (1999) Genetic engineering of recombinant glycoproteins and the glycosylation pathway in mammalian host cells. *Glycoconjugate J* **16**: 81–97.

Grabowski GA, Barton NW, Pastores G *et al.* (1995) Enzyme therapy in type 1

Gaucher disease: comparative efficacy of mannose-terminated glucocerebrosidase from natural and recombinant sources. *Ann Internal Med* **122**: 33–39.

Hamilton SR, Bobrowicz P, Bobrowicz B *et al.* (2003) Production of complex human glycoproteins in yeast. *Science* **301**: 1244–1246.

Hamilton SR, Hopkins D, Wischnewski H *et al.* (2006) Humanization of yeast to produce complex terminally sialylated glycoproteins.

Harter C and Wieland F (1996) The secretory pathway: Mechanisms of protein sorting and transport. *Biochim Biophys Acta Rev Biomembrane* **1286**: 75–93.

Helenius A and Aebi M (2001) Intracellular functions of N-linked glycans. *Science* **291**: 2364–2369.

Herscovics A (1999) Importance of glycosidases in mammalian glycoprotein biosynthesis. *Biochim Biophys Acta Gen Subj* **1473**: 96–107.

Herscovics A (2001) Structure and function of Class I alpha 1,2-mannosidases involved in glycoprotein synthesis and endoplasmic reticulum quality control. *Biochimie* **83**: 757–762.

Herscovics A and Orlean P (1993) Glycoprotein-biosynthesis in yeast. *FASEB J* **7**: 540–550.

Higuchi M, Oh-Eda M, Kuboniwa H, Tomonoh K, Shimonaka Y and Ochi N (1992) Role of sugar chains in the expression of the biological activity of human erythropoietin. *J Biol Chem* **267**: 7703–7709.

Kalsner I, Hintz W, Reid LS and Schachter H (1995) Insertion into *Aspergillus nidulans* of functional Udp-Glcnac: α3-D-mannoside β-1,2-N-acetylglucosaminyl-transferase I, the enzyme catalyzing the first committed step from oligomannose to hybrid and complex N-glycans. *Glycoconjugate J* **12**: 360–370.

Kim YK, Shin HS, Tomiya N, Lee YC, Betenbaugh MJ and Cha HJ (2005) Production and N-glycan analysis of secreted human erythropoietin glycoprotein in stably transfected Drosophila S2 cells. *Biotechnol Bioeng* **92**: 452–461.

Kornfeld R and Kornfeld S (1985) Assembly of asparagine-linked oligosaccharides. *Annu Rev Biochem* **54**: 239–257.

Krezdorn CH, Kleene RB, Watzele M, Ivanov SX, Hokke CH, Kamerling JP and Berger EG (1994) Human beta-1,4 galactosyltransferase and alpha-2,6 sialyltransferase expressed in *Saccharomyces cerevisiae* are retained as active enzymes in the endoplasmic reticulum. *Eur J Biochem* **220**: 809–817.

Kukuruzinska MA and Lennon K (1998) Protein N-glycosylation: molecular genetics and functional significance. *Crit Rev Oral Biol Med* **9**: 415–448.

Lehle L and Bause E (1984) Primary structural requirements for N-glycosylation and O-glycosylation of yeast mannoproteins. *Biochim Biophys Acta* **799**: 246–251.

Lewis MJ, Sweet DJ and Pelham HRB (1990) The Erd2 gene determines the specificity of the luminal Er protein retention system. *Cell* **61**: 1359–1363.

Li H, Sethuraman N, Stadheim TA *et al.* (2006) Optimization of humanized IgGs in glycoengineered *Pichia pastoris*. *Nat Biotechnol* **24**: 210–215.

MacDougall IC (2002) Optimizing the use of erythropoietic agents – pharmacokinetic and pharmacodynamic considerations. *Nephrol Dialysis Transplant* **17**: 66–70.

Malissard M, Zeng S and Berger EG (1999) The yeast expression system for recombinant glycosyltransferases. *Glycoconjugate J* **16**: 125–139.

Malissard M, Zeng S and Berger EG (2000) Expression of functional soluble forms of human beta-1,4- galactosyltransferase I, alpha-2,6-sialyltransferase, and alpha-1,3-fucosyltransferase VI in the methylotrophic yeast *Pichia pastoris*. *Biochem Biophys Res Commun* **267**: 169–173.

Maras M and Contreras R (1994) *Methods of Modifying Carbohydrate Moieties*. United States, Alko Group Ltd, Helsinki, Finland.

Maras M, Saelens X, Laroy W, Piens K, Claeyssens M, Fiers W and Contreras R (1997) In vitro conversion of the carbohydrate moiety of fungal glycoproteins to mammalian-type oligosaccharides – Evidence for N-acetylglucosaminyltrans-ferase-I-accepting glycans from *Trichoderma reesei*. *Eur J Biochem* **249**: 701–707.

Martinet W, Maras M, Saelens X, Jou WM and Contreras R (1998) Modification of the protein glycosylation pathway in the methylotrophic yeast, *Pichia pastoris*. *Biotechnol Lett* **20**: 1171–1177.

Maynard Y and Baenziger J (1981) Oligosaccharide specific endocytosis by isolated rate hepatic reticuloendothelial cells. *J Biol Chem* **256**: 8063–8068.

McAleer WJ, Buynak EB, Maigetter RZ, Wampler DE, Miller WJ and Hilleman MR. (1984) Human hepatitis-B vaccine from recombinant yeast. *Nature* **307**: 178–180.

Mistry PK, Wraight EP and Cox TM (1996) Therapeutic delivery of proteins to macrophages: implications for treatment of Gaucher's disease. *Lancet* **348**: 1555–1559.

Moremen KW and Robbins PW (1991) Isolation, characterization, and expression of cDNAs encoding murine alpha-mannosidase-II, a Golgi enzyme that controls conversion of high mannose to complex N-glycans. *J Cell Biol* **115**: 1521–1534.

Moremen KW, Trimble RB and Herscovics A (1994) Glycosidases of the asparagine-linked oligosaccharide processing pathway. *Glycobiology* **4**: 113–125.

Munro S and Pelham HRB (1987) A C-terminal signal prevents secretion of luminal Er proteins. *Cell* **48**: 899–907.

Paulson JC and Colley KJ (1989) Glycosyltransferases – Structure, localization and control of cell type-specific glycosylation. *J Biol Chem* **264**: 17615–17618.

Pelham HRB, Hardwick KG and Lewis MJ (1988) Sorting of soluble Er proteins in yeast. *EMBO J* **7**: 1757–1762.

Rothman R, Perussia B, Herlyn D and Warren L (1989) Antibody-dependent cytotox-icity mediated by natural killer cells is enhanced by castanospermine-induced alterations of IgG glycosylation. *Mol Immunol* **26**: 1113–1123.

Roy SK, Yoko-O T, Ikenaga H and Jigami Y (1998) Functional evidence for UDP-galac-tose transporter in Saccharomyces cerevisiae through the in vivo galactosylation and in vitro transport assay. *J Biol Chem* **273**: 2583–2590.

Schachter H (1991) The 'yellow brick road' to branched to complex N-glycans. *Glycobiology* **1**: 453–461.

Schachter H (2000) The joys of HexNAc. The synthesis and function of N- and O-glycan branches. *Glycoconjugate J* **17**: 465–483.

Schwientek T and Ernst JF (1994) Efficient intra- and extracellular production of human beta-1,4-galactosyltransferase in *Saccharomyces cerevisiae* is mediated by yeast secretion leaders. *Gene* **145**: 299–303.

Schwientek T, Narimatsu H and Ernst JF (1996) Golgi localization and in vivo activ-ity of a mammalian glycosyltransferase (human beta1,4-galactosyltransferase) in yeast. *J Biol Chem* **271**: 3398–3405.

Semenza JC, Hardwick KG, Dean N and Pelham HRB (1990) Erd2, a yeast gene required for the receptor-mediated retrieval of luminal Er proteins from the secretory pathway. *Cell* **61**: 1349–1357.

Sharma S, Knauer R and Lehle L (2001) Biosynthesis of lipid-linked oligosaccharides in yeast: the ALG3 gene encodes the Dol-P-Man:Man5GlcNAc2-PP-Dol manno-syltransferase. *Biol Chem* **382**: 321–328.

Stockert R (1995) The asialoglycoprotein receptor: relationships between structure, function and expression. *Physiol Rev* **75**: 591–609.

Strahl-Bolsinger S, Gentzsch M and Tanner W (1999) Protein O-mannosylation. *Biochim Biophys Acta Gen Subj* **1426**: 297–307.

Tabas I and Kornfeld S (1979) Purification and characterization of a rat-liver Golgi alpha-mannosidase capable of processing asparagine-linked oligosaccharides. *J Biol Chem* **254**: 1655–1663.

Teasdale RD and Jackson MR (1996) Signal-mediated sorting of membrane proteins between the endoplasmic reticulum and the Golgi apparatus. *Annu Rev Cell Dev Biol* **12**: 27–54.

Thibault P (2001) Identification of the carbohydrate moieties and glycosylation motifs in *Campylobacter jejuni* flagellin. *J Biol Chem* **276**: 34862–34870.

Tulsiani DRP, Hubbard SC, Robbins PW and Touster O (1982) Alpha-D-mannosidases of rat-liver Golgi membranes – Mannosidase-Ii is the Glcnacman5-Cleaving enzyme in glycoprotein-biosynthesis and mannosidase-Ia and mannosidase-Ib are the enzymes converting Man9 precursors to Man5 intermediates. *J Biol Chem* **257**: 3660–3668.

Van Vliet C, Thomas EC, Merino-Trigo A, Teasdale RD and Gleeson PA (2003) Intracellular sorting and transport of proteins. *Prog Biophys Mol Biol* **83**: 1–45.

Vervecken W, Kaigorodov V, Callewaert N, Geysens S, De Vusser K and Contreras R (2004) In vivo synthesis of mammalian-like, hybrid-type N-glycans in *Pichia pastoris*. *Appl Environ Microbiol* **70**: 2639–2646.

Vinogradov E, Petersen BO, and Duus JO (2000) Isolation and characterization of non-labeled and 13C-labeled mannans from Pichia pastoris yeast. *Carbohydr Res* **325**: 216–221.

Walsh G (2003) Biopharmaceutical benchmarks – 2003. *Nat Biotechnol* **21**: 865–870.

Ziegler FD, Gemmill TR and Trimble RB (1994) Glycoprotein-synthesis in yeast – Early events in N-linked oligosaccharide processing in *Schizosaccharomyces pombe*. *J Biol Chem* **269**: 12527–12535.

The Bioprocess

Perfusion or fed-batch? A matter of perspective

9

Marco Cacciuttolo

9.1 Introduction

The objective of this chapter is to provide an industry perspective on the selection criteria for using either the perfusion or the fed-batch techniques as the platform of choice to manufacture biologic products, made by cell culture and to provide an insight into the decision-making process in favor of perfusion-based manufacturing or fed-batch. The specific case described here corresponds to a multi-product facility for initial product development trials (i.e. phase I and phase II). This clarification of scope is important, as each technique is suitable for production of biologics, and the reasons to prefer one over the other as described in this chapter are primarily based on long-term objectives for the facility such as meeting the production requirements of a diverse pipeline rather than on the intrinsic merits of either technique.

Perfusion and fed-batch modes of bioreactor operation (illustrated in *Figure 9.1*) have been extensively used in the biotechnology industry for a number of years and are extremely popular for manufacturing the many different approved products (*Table 9.1*). The decision between using perfusion or fed-batch from a technical perspective could be a matter of viewpoints and individual preferences, and has been the subject of debates between the advocates of either technique. The objective of this chapter is to provide an industry perspective on the selection criteria for using either of these techniques. Ultimately, the decision has to do with long-term vision rather than short term goals.

The basic techniques for mammalian cell cultivation can be divided into batch and continuous mode of operation (Birch, 2000). Batch cultivation is perhaps the simplest way to operate a fermenter or bioreactor. It is easy to scale up, easy to operate, and offers a quick turn around and a reliable performance. Vessel sizes of 20 000 L have been reported for animal cell cultivation and of over 100 000 L for fermentation are also available. The major limitation of a batch is the accumulation of toxic metabolites and the depletion of nutrients. This is resolved by the use of bolus or continuous feeds (typically referred to as fed-batch). The success of a fed-batch culture depends on carefully timing the feeds and exquisite control of osmolality versus nutritional demands. Today, it is not unusual to reach well over 5 g L^{-1} of recombinant antibody concentrations using this technique (Cacciuttolo *et al.*, 2006). Continuous processes can be classified as cell

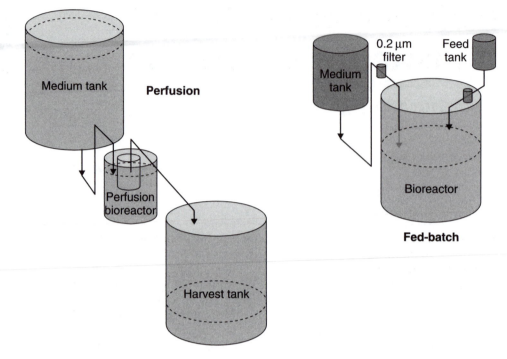

Figure 9.1

Technologies for mass production. Perfusion and fed-batch.

Table 9.1 Examples of production methods for biotechnology-derived products

Protein	Clinical application	Production process
Recombinate (F VIII)	Hemophilia	rCHO, bleed-feed
Kogenate (F VIII)	Hemophilia	rBHK-21, bleed-feed
Pulmozyme (Dnase I)	Cystic fibrosis	rCHO, suspension
Cerezyme	Gaucher's disease	rCHO, microcarriers
Activase (tPA)	Thrombolytic agent	rCHO, suspension
Epogen (Epo)	Stimulation of erythropoiesis	rCHO, Roller bottles
Rituxan (Mab)	B-cell nonHodgkin's lymphoma	rCHO, fed-batch
Synagis (Mab)	Prevention of RSV disease	rNS/0, fed-batch
Herceptin (Mab)	Breast cancer	rCHO, suspension
OKT3 (Mab)	Rescue of acute renal rejection/GVHD	Mouse ascites
Zenapax (Mab)	Prevention of acute renal rejection	rNS/0, suspension
Reopro (Mab)	Prevention of cardiac ischemic complications	rSP2/0, perfusion
Remicade (Mab)	Reumatoid arthritis	rSP2/0, perfusion
Enbrel	Rheumatoid arthritis	rCHO
Avastin (Mab)	Metastatic colorectal cancer	rCHO
Erbitux (Mab)	Metastatic colorectal cancer	SP2/0, fed-batch
Humira (Mab)	Reumathoid arthritis	rCHO, fed-batch
Xolair(Mab)	Metastatic colorectal cancer	rCHO, fed-batch
GenHevac B (HbsAg)	Hep B vaccine	rCHO,microcarriers
FluMist	Influenza virus vaccine	Eggs

retention or non-cell retention. Perfusion is perhaps the most popular method of continuous cultivation for animal cells. It is claimed to sustain high productivity for months of continuous operation (Hess, 2004). The devices typically used for cell retention are spin filters, hollow fibers, and decanters. Large-scale operation of perfusion processes can reach up to 2000 L of bioreactor working volume. Typically, the process is operated at one to two bioreactor volumes exchanged per day. Perfusion is one variation of a continuous process in which cells are retained within the bioreactor to achieve the highest level of cell density possible (Deo *et al.*, 1996). Usually, high productivity in cell culture is achieved by a high specific productivity and/or high cell density. Perfusion processes also offer the advantage of minimizing the size of the bioreactor and the 'down time' of the production units, and homogeneity of product quality throughout the production cycle as cells are kept in a physiological steady state.

Given the choices, emerging biotechnology companies need to carefully allocate their financial resources to the areas that will ultimately determine their success. The strategy in most cases is to conserve precious cash and invest in product discovery and clinical research. This means that manufacturing decisions are focused on keeping the product development expenses to a minimum and on having expedient process development timelines. Scale-up is usually deferred to the last possible moment, usually to the point where there is enough evidence of product activity to justify the expenditure, or when additional cash is available to afford a push forward at risk in case there is a race against a competitor. In this scenario, there are no second chances and margin for errors, and all aspects of the decision-making process need to be carefully scrutinized and weighed. In this regard, the technology platform chosen early on in the development of a given product could be of critical importance in the overall and long-term strategy of a given company. This is because the decision on which technology platform is long-lasting and determines all future investment decisions.

9.2 Factors affecting the decision on choosing the manufacturing technology

9.2.1 Technology expertise

Several areas of recent development are of importance when choosing one technology over the other for long-term use. For instance, it is well known that product titers in cell culture today are at least 1 g L^{-1}. With continued improvements, the trend shows that 10 g L^{-1} will be reached within the next few years (see *Figure 9.2*). These improvements are due to sophisticated feeding techniques and improved clone selection procedures, which combined can generate high producing cell lines in a relatively very short time (4–6 months). These improvements make it possible for fed-batch to compete with perfusion-based processes in terms of overall throughput, with lower expense (as discussed later in this chapter), and with greater flexibility.

The most obvious and fundamental difference between a fed-batch and a perfusion process is the duration of the batch. It is well understood that

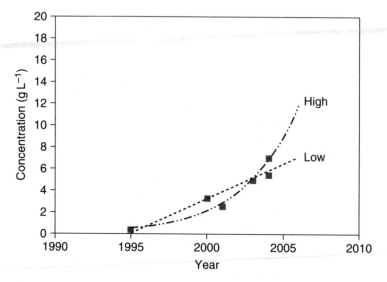

Figure 9.2

Projections for antibody concentrations obtained from cell culture (Cacciuttolo, 2006).

maximizing the productive time of a bioreactor will determine the overall economics of the process. From that view point, perfusion processes offer an advantage over fed-batch, as it only requires one start up for continuous production over a period of months. An equivalent fed-batch process in terms of overall throughput and assuming a five times larger bioreactor, will require several starts and down time, reducing its overall economy. However, this analysis does not take into account the fact that personnel requirements are much higher to sustain continuous production. This makes the comparison between perfusion and fed-batch, much closer in terms of costs per gram when all of these items are carefully considered and balanced (*Table 9.2*).

The availability of contract manufacturers for either technology is perhaps the single most important aspect of the decision making process in favor of either technology for an emerging biotechnology company (Koberstein, 2005). This is because large-scale manufacturing capability has to be outsourced by small- to medium-size biotechnology companies in most cases. Having a large pool of fed-batch-based CMOs to choose from, helps tremendously in regards to availability, costs, expertise, and regulatory track record.

Another important aspect for a product pipeline rich company is to have the ability to make products using processes of short duration. This allows the manufacturer the flexibility of accommodating many different products in a given facility. As *Table 9.3* illustrates, it is possible to make three times more different products per year per bioreactor unit when using a fed-batch approach rather than perfusion.

Fed-batch is also a viable strategy to increase the ability to manufacture many different products at early stages of pipeline development within the same facility. This is particularly important for companies having a rich pipeline of products. However, high titers in the bioreactor (i.e. over

Table 9.2 Comparison between perfusion and fed-batch processes

Perfusion	Fed-batch
Requires few starts	Multiple starts for equivalent throughput
Amenable for labile products	Robust products are best suited
Homogenous product quality throughout	Consistent product. It requires careful selection of harvest time
Long cycle times: minimizes downtime	Fast turn around: minimizes development and validation time
Needs to define the meaning of a lot	Each discrete run is one lot
Continuous need to make medium, process harvests, and remove waste	Discrete steps
Requires robust automation and monitoring of equipment (liquid level, medium and harvest flow, etc)	Minimum requirements for automation
Requires highly specialized technical expertise	Know-how widely available
Contamination events may affect multiple harvests within a run	Contamination events affect only specific lots
Requires high number of operators	Requires relatively low number of operators
Requires high need for instrumentation and control	Low requirements for instrumentation and control
Requires large footprint	Requires relatively small footprint
High volumes of culture medium	Low use of culture medium for same throughput
Technology for cell retention is usually proprietary and it requires specialized equipment	Fed-batch does not required additional and/or specialized equipment
Requires constant monitoring	Requires sporadic monitoring
Larger costs for waste handling and disposal	Lower costs for waste handling and disposal
Higher process contaminants/product ratio	Lower process contaminant/product ratio
Long validation timelines	Short validation timelines
Complex technology transfer	Simple and fast technology transfer
Product is more diluted	Product is more concentrated in the harvest
Few qualified and expert CMOs available	Large selection of capable and licensed CMOs
Short residence times (ideal of labile products)	

Table 9.3 Bioreactor cycle-times

Perfusion	Fed-batch
Ramp-up time: 5 days	Ramp-up time: Not applicable
Production time: 40 days	Production time: 14 days
Turn-around time: 5 days	Turn-around time: 2 days
Total cycle time: 50 days/run	Total cycle time: 16 days/run
Number of runs/year/biorx: 300/50 = 6	Number of runs/year/biorx: 300/16 = 18

300–500 mg L^{-1}) are key to making the fed-batch approach feasible. Therefore, efforts at increasing cell line expression levels are essential and having a strong process development group is a precious resource. Today, due to the focus on this area by many industry leaders, multi-gram per liter titers are standard (see *Figure 9.2*), and are usually obtained within a few months from the transfection step.

The size of the bioreactor is important in assessing total throughput. This is illustrated in a comparison between the two techniques (*Table 9.4*). Using

Table 9.4 Throughput considerations between fed-batch and perfusion

Perfusion	Fed-batch
Product titer (P)	
Perfusion rate (PR): 1 reactor volume per day	Maximum cell density (X_M): 6×10^6 cells mL^{-1}
Cell density at the steady state (X_{SS}): 20×10^6 cells mL^{-1}	Integral of cells : 50.5×10^6 cells mL^{-1} (over a 14-day period)
Product concentration in the harvests at the steady state (P_{pf}): $P_{pf} = (X_{SS} \times q_p)/PR = 400$ mg L^{-1}	P_{fb} (xdt) $\times q_p = 1.01$ g L^{-1}
(Note: These values match actual experimental results)	
Overall productivity (Q_p)	
PR or 1 RV/day = 100 L day^{-1}	
Production time (T): 40 days/run (from *Table 9.3*)	
So, using values given in *Table 9.3*:	
$Q_p = PR \times P_{pf} \times T$	$Q_p = P_{fb} \times RV = 1.01$ g L$^{-1} \times 100$ L
$\quad = 40$ g day$^{-1} \times 40$ days/run	$\quad = 100$ g/run
$\quad = 16$ kg/run $\times 6$ rums/year/biorx	$\quad = 100$ g/run $\times 18$ runs/year/biorx
$\quad = 9.6$ kg year^{-1} per biorx	$\quad = 1.8$ kg year^{-1} per biorx

The calculations assume a cell line with a productivity (q_p) of 20 pg protein cell^{-1} day^{-1}, and a bioreactor size of 100 L.

the same cell line, it clearly shows what has been the standard conclusion: that a smaller bioreactor can be used for perfusion than in a fed-batch mode for the same throughput. This has been one of the main arguments to justify perfusion processes, as it is perceived that the cost of the bioreactor unit is the substantial investment of a manufacturing facility. However, media and harvest tanks are actually about 5–10 times larger in perfusion processes and occupy a large footprint (*Figure 9.1*). In fact, dedicated media and harvest tanks are not needed when using the fed-batch approach. Therefore, the benefit of having a small bioreactor size is marginal in the overall comparison, especially when the additional costs of cleaning and the validation of media and harvest tanks are included.

It is important to notice that the product concentration in the harvest of a fed-batch process is higher than in a perfusion process, using the same cell line. It is an important feature in several ways. A more concentrated product harvest does not require a concentration step typically used to reduce the volume of product handled in recovery and purification operations. However, the most striking benefit of a more concentrated harvest is that the ratio of product to process contaminants, such as DNA, host cell protein, growth factors, is much higher in a fed-batch process. This has a tremendous positive effect on product recovery and purification steps. Overall process yields increase by 10% on average when switching from perfusion to fed-batch, as resin capacity is much better utilized with a larger ratio of product to contaminants in the capture step. Also, a higher margin of viral safety factor for the process may be observed for the fed-batch process.

In addition, some of the costs that are not obvious or are hidden are rarely clearly explained when making comparisons between these two techniques. The cost of waste handling and total footprint are relevant when designing a facility, and they only become clear when one has gone through the exercise and experience of performing both operations on a large scale.

Table 9.5 Comparison of cell culture medium consumption between perfusion and fed-batch for an equivalent product throughput of 9.6 kg year^{-1}

Perfusion	Fed-batch
Bioreactor volume: 100 L PR: 1 RV/day = 100 L day^{-1}	Bioreactor volume: 500 L (equiv. lot size)
Medium consumption (MC) PR × (T + ramp-up) 100 L day^{-1} × (40 + 5) days/run	

MC$_{pf}$ = 100 L day^{-1} × 45 days/run
    = 4500 L/run × 6 runs/year/biorx
    = 27000 L year^{-1} per biorx
	MC$_{fb}$ = 500 L/equiv. run × 8 runs/year
	    = 9000 L year^{-1} per equiv. biorx

Table 9.5 shows the comparison of medium consumption between the two techniques for a similar throughput, which is also an indication of the relative volumes of waste that need to be handled in either case. It clearly shows that perfusion uses three times more medium, and therefore requires three times more liquid handling capacity than a fed-batch process. The liquid handling capacity includes medium receiving, storage and holding, harvest storage, and also waste handling. In a GMP environment, all of these additional pieces of equipment and process steps have to be validated adding time and cost.

9.2.2 Facility design and scope (product dedicated versus multi-product)

Provided the technical expertise is broad and inclusive, the intended scope of the bioreactor facility is a major decision-making factor that determines the technology to use. That is, what is the primary goal for the facility? If the facility is meant to support a rich pipeline of products entering phase I, it is important that the manufacturing process needs to be short and a technology platform is in place to plug-and-play one product after another in the same facility, using the same equipment, with similar batch records and SOPs, and same training for the personnel. The product requirements to support phase I and II trials (*Table 9.6*) are relatively modest if the culture productivity is above 500 mg L^{-1}. If, however, the productivity is below 100 mg L^{-1}, a perfusion process is the only choice to provide enough product to compensate low titers with large volumes, or to dedicate a substantial proportion of the manufacturing capacity to a single product.

A typical fed-batch cell culture process lasts about 2 weeks in the bioreactor. Therefore, about 15 lots of new products per year per bioreactor can be

Table 9.6 Typical drug requirements for phase I and II trials

Phase of clinical trial	Typical drug requirement	Typical cell culture yields (fed-batch)	Harvest volume
Phase I Phase II	50–100 g/product 200–500 g/product	>300 mg L^{-1} >1.0 g L^{-1}	670 L/product 1000 L/product

expected. *Table 9.6* clearly shows that for these levels of drug requirements and cell culture yields, a facility of a combined lot size of 1000 L per lot can accommodate about 15 different products for phase I and II per year using a fed-batch approach. Here the important factor is not how much product per year can be produced in the facility, but rather how many different products can be made in the same facility per year.

If the technical expertise is in place in order to reach over 300 mg L^{-1} for early process development stage reliably, then the switch from perfusion to fed-batch becomes practical. In addition to the features described in *Table 9.2*, the switch is facilitated when:

(i)   convertible equipment (bioreactors) between perfusion and fed batch is available;
(ii)   skilled staff (expertise, training) is available;
(iii)   there is a drive to reduce direct costs (raw materials, FTEs, etc);
(iv)   there is a drive to reduce hidden costs (waste disposal, testing, space allocation, process control);
(v)   there is a need to develop a more efficient manufacturing strategy (amount of data, scientists' time, turn around, process validation time and effort);
(vi)   there is a need to facilitate scale-up and process robustness (scale-up factor, reproducibility, minimum additional training);
(vii)   there is a need to reduce time for process development and transfer to manufacturing;
(viii) there is a need to ease the transfer process to a CMO (expertise, time, cost).

9.3. Impact of switching from perfusion to fed-batch

9.3.1 Personnel requirements

Perfusion is intrinsically a labor intensive process because it is continuous. Fresh medium has to be constantly prepared and harvest has to be constantly collected and processed to re-utilize harvest tanks. To allow for seamless operation, harvest tanks per production bioreactor have to be multiple, increasing the cost of equipment and GMP space (footprint). Because of this, for a perfusion-based process, a dedicated crew for product recovery is usually required. This is completely eliminated in a fed-batch process, where the same personnel dedicated to bioreactor operations can also perform product recovery operations, as it is a 'stop-and-go' step (*Table 9.7*).

One advantage of switching from perfusion to fed-batch is that the additional training of personnel into the new technique is minimal. Most of the elements of the fed-batch mode of operation are in place. Batch records are much leaner and easier to follow, and the complicated standard operating procedures (SOPs) for assembly, operation, and monitoring of perfusion are completely eliminated. Also, tracking harvests and dealing with bioburden excursions of individual harvests may compromise the entire perfusion run. Therefore, a recovery from such a disastrous outcome is much more difficult and lengthy in perfusion than in a fed-batch case.

Table 9.7 Example of personnel requirements (it assumes one shift)

Perfusion	Fed-batch
Inoculum: 2	Inoculum: 2
Bioreactor crew: 4	Bioreactor crew: 3
Product recovery: 4	Product recovery: 0
Downstream purification: 4	Downstream purification: 4
QC support: 2	QC support: 1
TOTAL: 16	TOTAL: 10
Cycle time: 90 days per lot	Cycle time: 52 days per manufactured lot
Equivalent full time employee (FTE): $FTE_{pf} = 5.1$ FTE/lot/yr	$FTE_{fb} = 1.7$ FTE/lot/yr

In this example, the direct cost of labor is about 3 times higher for perfusion relative to a fed-batch process.
NOTE: 'Cycle time' includes purification and formulation steps to bulk product.

9.3.2 Liquid handling

Figure 9.3 illustrates the impact of switching from perfusion to fed-batch in terms of medium consumption, which is reduced by 90%. The consequent impact of this reduction of overall volume of cell culture generated is the reduction of harvest handling and storage, as well as on waste disposal and handling (also reduced by 90%).

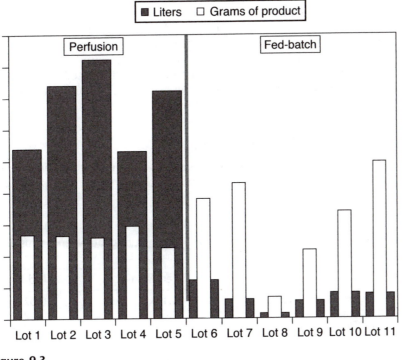

Figure 9.3

Impact on medium consumption of switching from perfusion to fed-batch.

The reduction of the volume of the waste to be handled after each lot is also a major benefit after switching from perfusion to fed-batch. It may actually take about 10 times longer to fill the waste collection tank. The elimination of frequent waste disposal operations obviously reduces its cost.

9.3.3 Equipment

In most cases, when using an internal spin filter as a cell retention device, the same perfusion bioreactors can be used in fed-batch process after minor alterations. The spin filters can be removed and impeller speeds adjusted, and re-calibrated relatively simply and quickly. The removal of external cell retention devices is even less complicated. This is very important in terms of reducing the cost and time of the overall project.

9.3.4 Manufacturing space

The reduction of cold storage needs for fresh medium and harvests can cause an immediate reduction of cold storage space by 55%. This space can be re-allocated to ambient processing and other functions within the facility. It actually allows the construction of parallel inoculum seed trains, which in turn increases the overall throughput of the facility by dramatically reducing product turn around time.

9.3.5 Decrease in cycle time

A fed-batch process can eliminate the need for product concentration, and reduce the overall time to manufacture and release of a lot from 130 days (for perfusion) down to 80 days. Bioreactor preparation time can be drastically reduced as the internal spin filter removal, disassembly, clean-out-of-place, reassembly, and testing can be completely eliminated. Those steps occupy a significant amount of time in the overall turn-around time for the preparation of bioreactor unit for the next lot.

9.3.6 Direct costs of manufacturing

A relative comparison of the cost drivers shows that fed-batch is about 34% less expensive than a perfusion process, for the same throughput (*Table 9.8*).

Table 9.8 Comparison of direct costs of manufacturing between perfusion and fed-batch

	Perfusion	Fed-batch
Cell culture medium utilization:	0.16×	0.07×
Waste handling:	0.05×	0.02×
FTEs (from *Table 9.7*)	0.6×	0.375×[a]
Other:	0.19×	0.19×
Cost per lot:	1× (basis)	0.66×

(a) (10/16) • 0.6×

9.3.7 Productivity and morale

Interestingly, one benefit of switching from perfusion-based mode to fed-batch, is that the scientists responsible for developing the manufacturing process can actually spend more time strategizing and generating multiple sets of data, rather than setting up and monitoring few long perfusion runs. These scientists can spend more time designing and creating new feed schemes, as opposed to changing medium and harvest containers, which ties them up on housekeeping activities. There is clear evidence of the requirement of more manpower just to maintain the perfusion operation running smoothly than fed-batch bioreactors. The recovery from unexpected data or technical failures in perfusion is much longer than when running a fed-batch process, causing more significant delays in technology transfer time to the manufacturing group. On the other hand, the manufacturing group which relies on its expertise on running perfusion technology may feel insecure about abandoning a hard-to-learn technology for a more standard and simpler one. Job security can be an issue and the staff need to learn additional techniques to make them more broadly proficient in cell culture.

When the technology platform is fed-batch, more time can be focused on creative thinking primarily in process development, organization, and training in GMP production, resulting in more projects that could be handled with the same personnel. In fact, amount of data for a single project increases substantially, as scientists spend more time on a single project. It also allows the scientists to develop platforms and templates that could easily be applied to new projects. Simpler technologies such as fed-batch are easier to transfer from the development to the manufacturing groups, either in-house or to a CMO, than the longer and more complex processes such as perfusion. However, one of the key factors that can make such a dramatic and momentous switch succeed or not is to have motivated staff. The people involved need to believe that it is the right thing to do.

9.4 Conclusions

Perfusion is a well established technology for producing large amounts of product. This technology is particularly suitable when product concentrations are low, and/or product is labile in the cell culture milieu. There are many companies that have decided to use of perfusion early on in their product development cycles due to those constraints. However, new advances in expression levels and media formulations are making the selection of fed-batch a much simpler method of culture and the platform of choice for many. Even for those companies that have traditionally used perfusion, the alternative of fed-batch is very attractive, and a complete switch from perfusion to fed-batch is a possibility (Koberstein, 2005). The main drivers for switching from perfusion to fed-batch are the availability of high producer cell lines, the larger number of available CMOs for fed-batch, the available development time per project, and a strong drive to decrease the overall operating costs per project. A facilitator for the switch from perfusion to fed-batch is that the bioreactors could be easily converted from perfusion to fed-batch; for instance, by simply removing the spin filter. Re-qualification work in this case is minimal, which in turn makes the batch records and SOPs to operate the bioreactors in fed-batch mode much simpler. The transformation from perfusion to fed-batch-based processes can be completed in about 2 years, and it needs to be managed carefully. For new projects, the switch is transparent and swift. However, for those products that are already in clinical trials when the switch is being proposed, one strategy is to make the cell line, medium, and production process changes (including bioreactor mode of operation and purification) all at once. The standard approach is to provide a comprehensive set of biochemical tests to show product comparability which may also include pharmacokinetic studies in animal and/or in humans. The ultimate justification for choosing perfusion versus fed-batch will be provided by a careful analysis of cost savings and efficiencies of one technology versus the other.

Acknowledgments

The author thanks Dr Pradip Ghosh-Dastidar for reviewing the manuscript and Beth Jacobs for its preparation.

References

Cacciuttolo M and Arunakumari A (2006) Scale-up considerations for biotechnology-derived products. In: *Pharmaceutical Process Scale-Up,* 2nd edn (M Levin, ed). Taylor & Francis, Boca Raton, FA, pp. 129–160.

Birch J (2000) Cell products – Antibodies. In: *Encyclopedia of Cell Technology,* (R. Spiers, ed).Wiley-Interscience, New York, pp. 415–419.

Deo YM, Mahadevan MD and Fuchs R (1996) Practical considerations in the operation and scale-up of spin filter based bioreactors for monoclonal antibody production. *Biotechnol Prog* 12: 57–64.

Hess P (2004) Using continuous perfusion cell-culture. Presented at biologic, Boston, MA, October.

Koberstein W (2005) Centocor enters center stage: Three business units play leading roles in biotech at Johnson & Johnson from *BioExecutive International,* October, p. 23.

Index